The High-Technology Connection:
Academic/Industrial Cooperation for Economic Growth

by Lynn G. Johnson

ASHE-ERIC Higher Education Research Report No. 6, 1984

Prepared by

ERIC® *Clearinghouse on Higher Education*
The George Washington University

Published by

Association for the Study of Higher Education

Jonathan D. Fife,
Series Editor

Cite as:
Johnson, Lynn G. *The High-Technology Connection: Academic/Industrial Cooperation for Economic Growth.* ASHE-ERIC Higher Education Research Report No. 6. Washington, D.C.: Association for the Study of Higher Education, 1984.

The ERIC Clearinghouse on Higher Education invites individuals to submit proposals for writing monographs for the Higher Education Research Report series. Proposals must include:

1. A detailed manuscript proposal of not more than five pages.
2. A 75-word summary to be used by several review committees for the initial screening and rating of each proposal.
3. A vita.
4. A writing sample.

Library of Congress Catalog Card Number: 84-73551
ISSN 0737-1292
ISBN 0-913317-15-2

ERIC® Clearinghouse on Higher Education
The George Washington University
One Dupont Circle, Suite 630
Washington, D.C. 20036

ASHE Association for the Study of Higher Education
One Dupont Circle, Suite 630
Washington, D.C. 20036

This publication was partially prepared with funding from the National Institute of Education, U.S. Department of Education, under contract no. 400-82-0011. The opinions expressed in this report do not necessarily reflect the positions or policies of NIE or the Department.

EXECUTIVE SUMMARY

The nation's recent recession has nurtured a new awareness of technology and its impact on global economic competition. Nowhere is this situation more evident than in the popular fascination with "high technology," a term that has been glamorized by the media, canonized by economic planners, and alternatively analyzed and puzzled over by academics, industrial executives, and others. Accompanying this rising consciousness of technology is a mounting concern for innovation in the private sector and increasing pressure for organizations of every description to act together in the revitalization of America's economy. Higher education and industry, in particular, are being called upon to form partnerships for technological progress in the public interest.

The purpose of this report is to examine the "high-technology connection" in which academic institutions and industrial firms are joined in cooperative efforts to stimulate technological change, taking into consideration the major spheres of academic/industrial cooperation, the primary mechanisms of interaction, and some of the most significant emerging problems and opportunities.

What Is the Context for Academic/Industrial Cooperation?
Economic planners increasingly regard academic institutions as critical resources in strategies to reinvigorate mature industries and stimulate new, "sunrise" industries. The basic research they perform, the skilled manpower they prepare, and the important services they provide are essential for corporations, entrepreneurs, and communities attempting to adapt to a rapidly changing economic and technological environment. Consider the following findings:

- The growth of thriving high-technology complexes, such as Boston's Route 128, California's Silicon Valley, and North Carolina's Research Triangle Park, owe much of their success to the concentration of intellectual capital in nearby universities.
- The attraction of new high-technology companies to any area depends in part on the responsiveness of community and technical colleges in offering training in technological skills.

- The ability of corporations to recruit competent professionals is enhanced by the availability of continuing education courses and other academic offerings at local institutions.

Surveys of state government initiatives reveal that over 200 programs related to high technology are currently in place. Many states are seeking to improve their educational infrastructure and to promote academic/industrial connections and thereby enhance their competitive position. Several governors are planning substantial increases in appropriations for higher education "to spur research and job training in high technology fields as an aid to economic development" (Magarrell 1984, p. 1).

What Makes up the "High-Technology Connection"?
From the standpoint of economic development, the high-technology connection can be conceptualized in terms of three major areas of linkage: research and development, technology transfer, and human resource development.

The research connection has received most attention because of advanced technology's dependence on fundamental research. Continuous innovation in products and processes requires close interaction between those who perform basic research and those who use it to develop and commercialize products (Prager and Omenn 1980, p. 379). Industries seek to gain a "window on technology" and access to outstanding faculty and graduate students. Universities need to supplement federal funding, secure adequate instrumentation, and support graduate research. This connection has some important features:

- Cooperative research projects often arise out of prior consulting relationships. Personal interactions and understanding of industry's needs are regarded as key factors in successful connections.
- Multimillion dollar agreements—such as the 12-year, $23 million compact between Harvard and Monsanto for biological and medical research—are very rare.
- Research connections include such mechanisms as contracted research projects, exchanges of personnel, and cooperative research centers. They vary in the extent of interaction, number of participating universi-

ties and industries, and degree of joint planning and governance.

Technology transfer is premised on the realization that the emergence of new concepts in science and engineering is only one part of the larger process of innovation. Analysis of innovation as it occurs in established corporations and in new firms reveals that numerous factors impinge on the process, including the costs and risks associated with research and development, the organizational practices of the corporation, the role of the entrepreneur as product champion or as head of a new firm, and the needs of small businesses for technical, managerial, and financial assistance.

- Transfer occurs through such means as informational events and publications, faculty consulting, associates programs, and various kinds of extension services.
- Industrial or research parks, often regarded as epitomizing campus/corporate ties, require the right mix of circumstances to succeed—and many have failed. A variation that is becoming more common is the industrial incubator that helps small firms in their initial, start-up phase.
- Efforts to support entrepreneurship often entail close cooperation with a number of industrial and community growth organizations in a particular area. The term "cooperative entrepreneurial development" describes these comprehensive programs.

The human resource connection merits more examination than it has received thus far. The supply of graduates in science and engineering and the availability of skilled technicians are already major concerns of technology-based industries (Joint Economic Committee 1982, p. 39; Useem 1981, pp. 19–20), and it is estimated that the skills of nearly half of the work force may be obsolete by the year 2000. To meet its need for workers, industry now spends from $10 billion to $30 billion or more annually on in-house training programs. This trend is not likely to change, but authorities in the training field see great opportunities for an expanded academic role, contingent on willingness to adapt to industry's needs (Lynton 1981, pp. 14–15).

- Mechanisms for linking academia with industry vary from regular degree offerings and industrial advisory committees to external degrees and programs that grant credit for noncollegiate learning. Sponsorship of continuing education courses and the cooperative education movement have grown dramatically.
- While four-year colleges and universities struggle with critical shortages of faculty in engineering and computer science, two-year institutions are trying to accommodate the new demands for technicians in advanced technology fields.
- Little statewide coordination of postsecondary education resources, specifically training and other needs related to technology, is apparent. Development strategies for the most part are lodged in departments of commerce or economic development.

Not many true "partnerships" exist in the fullest sense of joint planning and management or extensive interaction on many fronts, although some examples are emerging. Initiatives to bring all segments of the community together to address local and regional economic needs may prove to be powerful incentives for stronger bonds.

What Issues Affect Academic/Industrial Cooperation?
Interorganizational efforts are not easy to develop and manage successfully. Barriers include the different purposes of the academic and industrial sectors, constraints on time and other resources, and rigid organizational policies and rewards. Bridging the gap between sectors is a unique challenge that requires leadership capable of fostering communication and mutual trust (Peters and Fusfeld 1983, p. 42).

A number of perplexing issues arise at each point of the academic/industrial nexus. In research, academics consider intellectual property rights an issue, although industrial representatives tend to assume the issue can be resolved through negotiation. Nontraditional programs involving credit for extrainstitutional learning or tailor-made offerings are an issue in human resource connections, because they may be seen as diluting standards or giving up control of the curriculum. In entrepreneurial relationships, the financial ties of faculty inventors or of institutions themselves to new ventures pose thorny problems of conflict of interest.

Beyond these specific issues is the more general concern that industrial ties may subvert academic priorities and abridge academic principles. Academic/industrial relationships do have important implications for academic freedom, autonomy, and objectivity, but while these principles are important guidelines, they should not be viewed as absolutes:

> *If the university moves nearer to a partnership with industry, more resources can become available, but the university may relinquish some of its unique capabilities for unrestricted exploratory research and freedom of action. There are no absolutes, and the issues become matters of degree and common sense* (National Science Foundation 1982, p. 32).

Nevertheless, many academic men and women are likely to remain wary of such connections, and the issues will no doubt be the subject of continuing debate in the months ahead.

What Are the Implications of Cooperation between Academia and Industry?

Viewed as embracing entrepreneurial services and the development of human resources in addition to research, the high-technology connection implies roles for all types of colleges and universities. It is evident that many institutions across the country anticipate expanding their associations with industry. In many cases, the imperatives of regional economic development will shape these relationships in important ways.

- Industry cannot replace the federal government as the mainstay of academic science. Even the most optimistic projections assume that industrial support (now about 4 percent of the total) will remain a relatively small percentage of the income for academic research and development.
- Industry's training needs do not represent easy sources of income and enrollments. The necessity of adapting to industry's requirements will require new

thinking about the teaching/learning process and the means of ensuring quality.

- State funding for ambitious projects in technological development will not come without strings attached. Based on legitimate concerns for economic improvement, these new sources of support may tempt institutions to promise more than they can deliver. But the danger is great that academia will oversell what it can contribute.

Higher education and industry have a long and fruitful tradition of cooperation, and institutions wishing to strengthen links can build on past experience. It would be wise to assess areas where cooperative arrangements already exist, determine how well they are working, and suggest how they might be improved. Doing so will not necessarily lead to new positions or new policies, but in times such as these, higher education clearly needs to rethink its relationship with industry.

ASHE-ERIC HIGHER EDUCATION RESEARCH REPORT SERIES CONSULTING EDITORS

Lois S. Peters
Center for Science and Technology Policy
New York University

John M. Peterson
Director, Technology Planning
The B. F. Goodrich Company

Marianne Phelps
Assistant Provost for Affirmative Action
The George Washington University

Robert A. Scott
Director of Academic Affairs
State of Indiana Commission for Higher Education

Cliff Sjogren
Director of Admissions
University of Michigan

Al Smith
Assistant Director of the Institute of Higher Education &
Professor of Instructional Leadership & Support
University of Florida

Donald Williams
Professor of Higher Education
University of Washington

CONTENTS

FOREWORD

Since the establishment of the land-grant colleges and research universities, higher education has always had some interaction with industry. Most notably this has been the area of service (e.g., the agricultural extension agent and the farmer) and research (e.g., the development and technology transfer of the transistor). However, as a result of reduced traditional funding, this relationship is coming under new scrutiny. Today higher education more often looks toward business and industry as a source of funds, state-of-the-art equipment, and temporary faculty with up-to-date expertise in such fast-moving technological areas as computers.

Caution exists on both sides in developing this education-industry connection. Academe fears that a direct relationship with industry will negatively affect the free flow of knowledge, that there will be a tendency for faculty to "prostitute" themselves, and that the profit-making nature of the commercial sector will inhibit the traditional emphasis on pure research. On the business side, reluctance stems from the notion that academics still live in an ivory tower and do not cope well with the real world.

State and local government leaders, however, are encouraging a greater interaction between higher education and industry. They perceive such a relationship to be healthy for their economy. As government leaders witness the growth of such industrial areas as The Research Triangle in the Durham-Raleigh, N.C. area and Silicon Valley in southern California, they realize how important higher education institutions can be in stimulating the development of new high-technology industries.

As clearly reviewed in this report by Lynn G. Johnson, assistant provost and member of the graduate faculty in the field of higher education at the University of Akron, there are three major connections between higher education and industry:

1. Research - Cooperative arrangements can help stimulate faculty, productivity and discoveries.
2. Technology transfer - Faculty discoveries can be transferred directly to the commercial sector.
3. Human resource development - Industry's need for personnel training or re-training has never been greater.

It is probably correct that both higher education and industry should approach their partnerships cautiously, constantly reviewing how these connections may affect their primary missions. This report provides a better understanding of how academe and industry have worked together in the past and how they may benefit from partnerships in the future.

Jonathan D. Fife
Series Editor
Professor and Director
ERIC Clearinghouse on Higher Education
The George Washington University

ACKNOWLEDGMENTS

This work is dedicated to two great practitioners of the art of interorganizational relations whom I have been fortunate to have as teachers: to Henry W. Munroe, director of the New Hampshire College and University Council; and to W. Richard Wright, assistant to the president for off-campus relations at the University of Akron.

I wish to express appreciation to the following persons at the University of Akron for the support that made this study possible: Noel L. Leathers, provost emeritus; John Blough, head, Department of Educational Administration; and Frank Costa, director, Center for Urban Studies.

A number of persons were especially helpful in suggesting directions: John Peterson of B. F. Goodrich, Theodore Settle of NCR Corporation, William Toombs of Pennsylvania State University, and Harold Yiannaki of Youngstown State University. Anka Skrtic, with the assistance of Valerie Johnson, performed most of the bibliographic work, and Susan Beck, Ruth Conner, and Katherine Morrison prepared the manuscript.

THE HIGH-TECHNOLOGY CONNECTION

> *There is general agreement among economists and others that one of the most powerful forces influencing the American economy is technological change—the advance in knowledge relative to the industrial arts [that] permits . . . new methods of production, new designs for existing products, and entirely new products and services* (Mansfield 1968, p. 1).

Several forces have converged in recent years to create intense interest in academic/industrial relationships: (1) the urgent sense of global, economic competition; (2) the search for ways to stimulate technological progress; (3) the launching of ambitious technological development strategies by state governments and other agencies; (4) a new appreciation for the contributions of academic research, teaching, and service to economic development; and (5) a belief in the benefits of close cooperation between academic and industrial organizations.

Academic/ industrial cooperation is . . . a means to other ends, the chief of which is economic growth through advanced technology.

To speak of academic/industrial linkages as "the high-technology connection" is not to say the only reasons for advocating cooperation are technological or that no other substantive foci for interaction exist. But the most powerful pressures affecting these links today revolve around technological innovation, dissemination, and implementation. They are symbolized and even glamorized in the fascination with "high tech," and they are embodied in the myriad proposals for technological development that one encounters everywhere in community growth organizations, regional planning bodies, state development offices, and federal agencies. What the high-technology connection means, in short, is that academic/industrial cooperation is not an end in itself but a means to other ends, the chief of which is economic growth through advanced technology (Baer 1980; Low 1983).

This chapter deals with the contemporary economic context of collegiate/corporate relations. It has three purposes:

1. to examine high technology in the context of current industrial and economic problems;
2. to explore the connection between high technology and higher education; and

3. to outline state government initiatives that seek to link academic institutions and industry in the development of high technology.

The Economic Context

Current economic conditions and trends are the subject of considerable attention as the United States seeks to recover from the stagnation of the late 1970s and early 1980s. Several themes are apparent in the literature, all of which relate in one way or another to technology.

One theme is that America is losing ground to other industrial nations: "Our once unchallenged preeminence in commerce, industry, science, and technological innovation is being overtaken by competitors throughout the world" (National Commission on Excellence 1983, p. 11).

Indications of slippage in international competition are indeed numerous. During the last decade, the United States lagged behind Japan, France, Canada, Great Britain, and West Germany in expenditures for plants and equipment as a percentage of total output, and in 1980 America ranked lowest among its chief competitors in exports as a percentage of gross national product (Helms 1981, p. 3). With increases in productivity rates in West Germany, France, and Japan exceeding those of the United States in recent years, some fear that the overall productivities of these countries may surpass our own by the end of the present decade (Hewlett et al. 1982, p. 615).

A second theme is that fundamental changes are occurring in the bases of economic growth. Among the reasons typically cited for America's problems in international competition, the status of research and development figures prominently (Hansen 1983; Molitor 1981; Prager and Omenn 1980). If "knowledge and information industries are fast becoming decisive factors in the growth of the productive forces of nations," then funding for research and development (R&D), the number of graduates in science and high-technology disciplines, and the number of patents applied for and granted are among the most vital factors for economic progress in today's world (Molitor 1981, p. 24).

Studies of chemical, petroleum, electrical, and aircraft industries have demonstrated that R&D has a significant impact on rates of productivity increase (Hewlett et al.

FIGURE 1
R&D AND THE BALANCE OF TRADE: 1960–1978

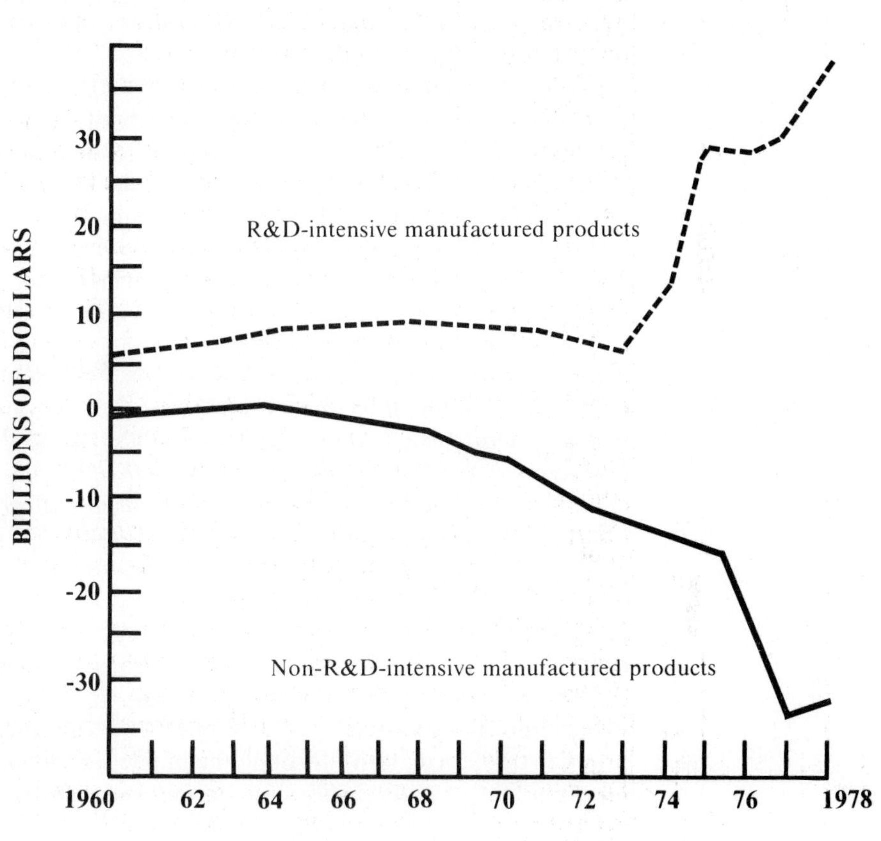

Source: Hewlett et al. 1982, p. 579

1982). Although it is difficult to isolate the effects of new technologies from investments in education and physical capital, it is clear that "these technologies play a major role in determining the size, viability, and profitability of particular industries and firms, as well as their competitiveness in international trade" (Hewlett et al. 1982, p. 578).

Striking evidence for the effects of research and development on worldwide competitive advantage can be seen in export/import data in various fields of manufacturing (see figure 1). Dramatic differences in the U.S. balance of trade (exports less imports) from 1960 to 1978 have been found in R&D-intensive and non-R&D-intensive manufactured product groups, with the former experiencing steady growth and the latter precipitous decline (Hewlett et al. 1982, pp. 578–79).

Congress recognized the importance of technological change for the national welfare when it created the Office of Technology Assessment in 1972:

> *As technology continues to change and expand rapidly, its applications are large and growing in scale, and increasingly extensive, pervasive, and critical in their impact, beneficial and adverse, on the natural and social environment* [Technology Assessment Act of 1972, §2, 42 U.S.C. §1862 (1972)].

Declarations of momentous economic change abound today, as various observers inform us that we are witnessing "a new socio-economic era" (Hansen 1983, p. 114), a shift to an "information society" (Naisbitt 1982, p. 20), and a "radical transformation" through the impact of technology on a host of occupations (National Commission on Excellence 1983, p. 11).

One far-reaching aspect of this change is the increasing interdependence of science and technology. Innovations in technology before the middle of the nineteenth century did not depend on scientific knowledge and were not introduced by trained scientists (Mansfield 1968, p. 2). But as scientific knowledge has developed, science and technology have increasingly interacted in many fields, leading in this century to whole new industries. Because science and technology are now closely intertwined, industrial firms that ignore R&D are risking obsolescence, and technological progress has become a matter of national concern (Birr 1966, pp. 76–77).

The third theme in the literature concerning the economic climate of the 1980s is the belief that advanced technology holds the key to future prosperity. Noting that numerous reports "state with confidence that new technology means increased wealth for people, companies, and countries," the president of the Exxon Research and Engineering Company suggests that industrial policy relating technological innovation to economic health is fast becoming "a key issue of our times" (David 1983, p. 27). While definitions of technology vary and assumptions about its impact are often questionable, it is clear that "the fashion for high tech . . . is more of a symptom than anything else. It means that the system is changing" (p. 30).

Odd as it may seem for such a ubiquitous term, "high technology" does not appear to have a universally accepted definition. One recent report uses "high technology," "new technology," and "advanced technology" interchangeably to refer to various forms of technological innovation, including changes within traditional industries as well as the emerging fields of microelectronics, telecommunications, and biotechnology (National Governors Association 1983, p. 4).

According to the *Dictionary of Business and Science,* technology is "the branch of knowledge that deals with the industrial arts; the unrestricted search for technological improvements" (Tver 1974, p. 522). While science involves knowledge of general truths or the operation of general laws, technology concerns "the applications of science to the needs of man and society" (Ashby 1958, p. 82). Derived from the Greek *techne,* meaning artistic or manual skill, technology is generally associated with those skills that result in a manufactured product or industrial process (Mansfield 1968, p. 1).

While some writers use high technology to refer to sophisticated products and processes—lasers, fiber optics, robotics, CAD/CAM, and the like—others prefer to limit the term to the knowledge that these products and processes embody. Its most important distinguishing characteristic does not lie in the technical content per se but in the "organization of knowledge and the continuum through which it is applied'' (Baker 1983, p. 111). The interdependence of system components and the integration of knowledge about these components in product design and manufacturing is therefore the hallmark of high technology (Baker 1983).

The fact that computers have aided in this integration and are used in many innovative products and processes no doubt explains the popular tendency to regard anything related to computers as high tech. But the essence of advanced technology would seem to be the ability to integrate and apply scientific and engineering knowledge to complex problems. In this sense, it is a relatively more sophisticated form of ''the know-how necessary for the creation of goods and services demanded by an economic society'' (National Academy of Sciences 1978, pp. 11–12).

One of the most widely read reports on technology and economic development identifies several industries in which the application of high technology is especially prevalent: chemicals and allied products; machinery, except electrical; electrical and electronic machinery, equipment, and supplies; transportation equipment; measuring, analyzing, and controlling instruments, including photographic, medical, and optical goods, and watches and clocks (Joint Economic Committee 1982, p. 4). Based on its survey of 691 companies in these categories, the Joint Economic Committee describes the high-technology firm as labor-intensive, science-based, and R&D-intensive. Such firms tend to be relatively young and relatively small, with a predominantly national or international market orientation (pp. 4, 19–21). Above all, they have high growth records and generate jobs—indeed, most of the new manufacturing jobs created in the private sector. From 1955 to 1979, these high-technology industries accounted for 75 percent of the net increase in manufacturing employment in the United States (p. 6).

The growth potential of products manufactured through advanced technologies—often referred to as ''high value–added products''—is frequently contrasted with the decline of traditional American industries and has led to their characterization as ''sunrise'' as opposed to ''sunset'' industries. The sunset industries, including such fields as automobiles, steel, and textiles, are expected to lose from 10 million to 15 million manufacturing jobs in the next 20 years, some of which will move to countries abroad where lower wages and access to raw materials offer competitive advantages (Georgia Office of Planning and Budget 1982, p. 9).

Given their growth potential, innovative features, and other desirable characteristics like employment of highly trained technical and professional people, high-technology companies are very attractive to communities (Hodges 1982; National Governors Association 1983)—hence the view that high technology holds the key to the revitalization of American industry and the future competitiveness of the American economy in worldwide markets.

> *High technology companies offer a brighter future for America, but they offer salvation for those regions of America that have borne the brunt of our economic decline. The ability of these states and localities to be a part of the technological renaissance will diversify their economies and make them less susceptible to large-scale economic downturns* (Joint Economic Committee 1982, p. v).

To regard high-technology companies as the salvation of whole regions, however, is probably overstating their potential. Even these firms are not immune to economic slumps, and in any case they are neither large enough nor numerous enough to replace all of the jobs being lost in declining industries (David 1983; Pollack 1984). A more sophisticated view is emerging: It recognizes the importance of new high-technology firms but takes more factors into account, including the continuing role of existing industries (Pollack 1984, p. 1). Thus, some professional planners now prefer the term "advanced technology" and define it broadly, emphasizing that technological innovation is crucial in all industrial sectors (Holtzman 1983).

Technology and Higher Education

The connection between technological development and higher education has several facets; the one most emphasized in the literature on academic/industrial connections is the research activity conducted on university campuses and its significance for industry. Universities conduct a relatively small part of the total R&D work in the United

States, but they are responsible for nearly half of all basic research—$4.3 billion or 49 percent in 1981 (Fusfeld 1983, p. 12).

Although basic research has traditionally been perceived as distinct from applied research and from development activity (see, for example, Pelz and Andrews 1966, p. 65), such distinctions are today becoming blurred and the time lag between discovery and commercialization compressed in many emerging fields (Culliton 1982). A recent essay on research occurring in medicine, recombinant DNA, energy, artificial intelligence, and other fields notes that the wedding of science and technology is everywhere apparent (Seitz et al. 1982). The evolution of computer capabilities and the development of new materials benefit nearly every field, and while the link between fundamental research and industrial use takes many forms, innumerable cases exemplify the tendency for breakthroughs in one area to spur advances in others.

Such observations have led to a number of generalizations about technological innovation that are germane to academic/industrial relations. In terms of the entire spectrum of activities from basic research to commercial application and marketing, the link between the generation of new knowledge and the translation of that knowledge into commercial products and services:

> *. . . depends on close interaction between those who perform basic research and those for whom the results of basic research are the raw materials for product development and commercialization. . . . Strong university-industry relationships can enhance the basic research-innovation linkage* (Prager and Omenn 1980, p. 379).

"Development through technological innovation" is a central element in economic revitalization and is crucial to both the encouragement of small entrepreneurial businesses and the modernization of traditional industries (Holtzman 1983, pp. 2–3). The "innovation development cycle" is a series of stages: technology ideas; commercializable ideas; productization; marriage of technology with entrepreneur; start of a business; and expansion of a business. Technological development must address each stage of this cycle and must also address the entire process in a coordinated manner. Higher education is "a resource of singular importance," and academic participation is an important part of the process:

> *It has been generally recognized nationwide that if efforts to encourage development through technological innovation are to succeed, the university community can, and must, play a vital role* (Holtzman 1983, p. 4).

The most widely heralded instances of academic participation in technological development are the case histories of Boston's Route 128, California's Silicon Valley, and North Carolina's Research Triangle Park, in which the Massachusetts Institute of Technology, Stanford University, and the Triangle institutions (Duke, North Carolina State, and the University of North Carolina), respectively, played important roles. These well-known centers of high-technology industry began with a concentration of intellectual capital in scientific and engineering fields and became magnets generating industrial research projects, spawning new firms, drawing additional faculty and graduate students, and in turn attracting additional support from governmental and industrial sources (Birchfield 1982). Their development occurred over a considerable period of time and was aided by substantial financial investments.

No single element was responsible for the success of those areas.

> *Instead, a combination of factors, including research and teaching activities at great universities, a rich endowment of labor skills, venture capitalists, high technology entrepreneurs, and federal procurement activities in the area are intermingled to provide the intricate fabric of a "creative environment" that under[lies] the economic dynamics of the region* (Joint Economic Committee 1982, p. 49).

The academic connection nevertheless requires special attention. Citing an earlier study of 32 high-technology firms, the Joint Committee's report notes that "in an extraordinary number of cases a university played a major role in the companies' histories" (1982, p. 51). This role involved not only keeping company personnel informed of the latest research developments but also helping firms acquire competent technical staff.

This second aspect of higher education's connection with technological development, the provision of trained manpower, is highlighted in the Joint Committee's own analysis of high-technology companies and their preferences for relocation. The item high-technology firms most frequently rated as significant or very significant in selecting a region for relocation was labor skills and availability (Joint Economic Committee 1982, p. 23). The availability of technical workers—machinists, welders, computer programmers—was rated even higher than that of scientists and engineers, presumably because the latter's mobility makes it possible to recruit them from outside a region (pp. 24–25). Consequently, in emphasizing "the potential that universities have in economic development," especially in attracting high-technology companies, the report portrays this contribution in terms of supplying industry with trained personnel as well as providing basic research (p. 39).

A third way in which colleges and universities are now being perceived as important resources for technological development is in the provision of entrepreneurial services. Many academic institutions are becoming involved "in a wide range of activities to stimulate the new business development process and entrepreneurial activity in their region" (*Venture Capital Journal* 1983, p. 7). In addition to their research and educational contributions, these institutions are providing such services as incubator units and science parks, technical and management assistance to entrepreneurs, and even direct or indirect investing.

The development of high-technology enterprise depends on many factors. The necessary ingredients for a "fertile business development climate" include the availability of venture capital; a fiscal, regulatory, and political environment supportive of the development of new enterprise; a business community familiar with the problems of rapidly growing young companies; major corporate or government R&D centers; a skilled labor force; technical and professional services; and commercial banks experienced in lending to nontraditional firms (*Venture Capital Journal* 1983, p. 7). Some institutions are attempting to bring all of these factors together or to act as a catalyst for bringing them together as part of an overall strategy for development.

State Planning for Technological Development

Given the perception that technology is critical for economic growth and that academic institutions are important resources for technological innovation, it is not surprising that economic planners now attach considerable significance to academic participation in economic development and to academic cooperation with the private sector. These connections are, of course, expressed in various ways. In a number of cases, it is a question of creating the right climate, environment, or set of conditions, and strong links between the academic and business communities and local government are a critical part of a climate conducive to technological development (Hodges 1982, pp. 4–5).

According to a survey by the National Governors Association (1983), state initiatives fall primarily into five categories: (1) policy development; (2) economic incentives; (3) technical support for businesses; (4) worker training; and (5) industry/university linkages. Academic involvement is evident in all of these categories, and examples of higher education's cooperation with business and industry abound. "University-industrial relationships are an essential component of state activities to encourage technological innovation," and when such state initiatives are added to efforts to improve mathematics and science education and to provide more reliable labor market information, they can further the goals of economic development. It is the sum total of such efforts that constitutes a comprehensive program: "All of these elements, taken together, should form a strong foundation for technological innovation processes in the states" (National Governors Association 1983, p. 14).

Another inventory of state programs, that conducted by the Office of Technology Assessment (1983), uses a taxonomy based on five types of programs (high-technology development, high-technology education, capital assistance, labor/technical assistance, and general industrial development) and 40 types of services. The report identifies over 200 programs with at least some features of high-technology development, although fewer states have a specific program or agency charged with promoting high technology—about 38 programs in 22 states. These latter programs focus on the needs of technology-based firms for technical, manpower, business, and financial assistance (Office of Technology Assessment 1983, p. 1).

High-technology initiatives are often extensions of overall economic development programs, and variations in developmental strategies reflect the different circumstances of each state. States with an existing research base seek to strengthen and retain the high-technology industry already there. States heavily dependent on sunset industries usually emphasize diversification and the application of new technologies to traditional manufacturing processes. Less industrialized states are likely to target expanding high-technology firms, hoping to attract new production facilities (Office of Technology Assessment 1983, p. 8).

The planning documents issued by development agencies in the various states point up common threads and differences in state programs. Georgia, for example, is emphasizing the attraction and creation of high-technology companies (Georgia Office of Planning and Budget 1982). The strategy is to identify key growth industries and to seek to create industry clusters or complexes of similar firms. This approach is contrasted with conventional methods of sending trade missions to woo individual companies and then providing assistance with site location.

Ohio has launched the Thomas Alva Edison Partnership Program "to foster cooperative research and development efforts involving enterprises and educational institutions that will lead to the creation of jobs" (Ohio Revised Code §122.33(c), 1983). With an appropriation of $32.4 million from the Ohio General Assembly, the Edison program includes as its major components "innovative research financing," which encourages smaller (up to $250,000) applied research projects, and "advanced technology application centers," which represent an effort to create four to six "world class technology research, development, and implementation centers" (Ohio Department of Development 1983, p. i).

Advanced technology centers are also a key element in Pennsylvania's development strategy. The Ben Franklin Partnership Program, actually initiated about a year before Ohio's Edison program, has already established centers at Lehigh University, at the University City Science Center in Philadelphia, in Pittsburgh, and at Pennsylvania State University (Pennsylvania Department of Commerce 1983). In each case, several academic institutions, numerous business corporations, and other organizations are involved in joint research and development efforts. Like the program in Ohio, state allocations must be matched by private funds, and a high degree of academic/industrial collaboration is required. Pennsylvania's centers, however, each involve a number of different areas of technology, while each of Ohio's centers is intended to have a single substantive R&D focus.

These economic development initiatives and the intense competition among the states for technological advance are apparent in several states.

> *In many states . . . there are concerted efforts to recruit and to build a high level technology industrial future, with considerable emphasis on the potential contributions of the research universities. Many of our most prominent scientists have become identified with these efforts* (Omenn 1983, p. 21).

Although higher education's high-tech connection has several facets as do states' efforts to encourage academic/industrial relationships, a great deal of the literature emphasizes the R&D connection, especially the contributions of the leading research universities. Accordingly, cooperation in research and development is an appropriate starting point for discussion of academic/industrial relationships in general. As the nature of research interactions becomes clear, some of the broader issues of technological innovation alluded to previously will also take on sharper focus, and the importance of other types of linkage will become evident as well.

Research is an essential component of the innovation process. Basic research may be the most important because it begins with seminal questions and a broad spectrum of possible explanations. To the extent that universities and technology-oriented corporations find ways to interact more effectively on basic scientific research, the usefulness of basic research information will also increase. This conclusion does not mean diversion of university researchers by industry. Rather, it implies mutual recognition of the value of applicable ideas or the feedback which often comes from application of relevant basic ideas (National Commission on Research 1980, p. 25).

'he research enterprise in America is highly pluralistic, ⁄ith multiple performers and multiple sources of financing. ‡esearch is performed in colleges and universities, in non- ·rofit research institutes and hospitals, in industrial labora- ɔries, and in governmental laboratories. Support derives ɾom a host of federal agencies, state governments, munici- ·alities, private foundations, voluntary public giving, and ıdustry (Seitz and Handler 1982).

The research enterprise in America is highly pluralistic, with multiple performers and multiple sources of financing.

Cross-sector connections for the performance and sup- ·ort of research are also numerous and complex. In the Jnited States, academic/industrial research ties in particu- ır are marked by a diversity and a vitality that are unpar- lleled in the world (Thackray 1983). This chapter de- cribes historical roles in R&D, incentives for greater ooperation, and the major types of research cooperation ·etween higher education and industry.

‡esearch Ties in Historical Perspective

'rom a historical perspective, the growing interest in cademic/industrial cooperation in research that has merged in the last several years merely brings to the ɔrefront relationships that have developed over many ecades.

In the nineteenth century, a vision of academe's poten- al research, educational, and service contributions to the ommunity was the basis for federal legislation focused on griculture and the mechanical arts. The Morrill Act of 862 established the land grant colleges, the Hatch Act of 887 added stations for agricultural research in the states, nd the Smith-Lever Act of 1912 funded agricultural extension work, "thus completing the integration of research, education, and technology transfer which constitutes the agricultural model of university-industry cooperation" (General Accounting Office 1983, p. 35).

Many of today's forms of cooperation in research and in the spread of technology can also be traced to the beginning of the century or earlier (Thackray 1983). University faculty practiced industrial consulting, at least occasionally, even before 1900. Industrial grants, especially to centers and institutes involved in applied research and service were being made to MIT and other institutions in the first decades of this century. Fellowship awards were part of the program at Pittsburgh's Mellon Institute, founded in 1913. Industrial associates programs and other extension services similarly had their origins before World War I. Thus, by 1914, "universities and industry were already closely linked on many levels" (Thackray 1983, p. 216).

Research within industry itself has exhibited phenomenal growth since General Electric built the first major industrial laboratory in the United States in 1900 (Jefferson 1982, p. 260). By 1920, 300 industrial laboratories were in operation—by 1930, 1,625 of them (Dupree 1957, p. 337). From 1921 to 1950, the number of industrial researchers grew at a rate of 9.6 percent annually (Thackray 1983, p. 206). Focused initially in the electrical and chemical fields, research and development has gradually diffused throughout the private sector, though it is still distributed unevenly among industries and the greatest portion of it occurs in the large laboratories of major corporations (Birr 1966, pp. 69–70). In major companies today, especially in technology intensive industries, one finds laboratories and staffs "that are the equal of or superior to those found in universities" (Shapero 1979, p. 7).

World War II brought industry, academia, and government together at an unprecedented level of cooperation for war-related research. "All the estates of science were drawn into a single great effort of applied science" (Dupree 1957, p. 373), with profound implications for future relationships. The wartime research effort enlarged the scale of activity, increased the number and intensity of contacts between academics and industrialists, and yielded spectacular results that "made plain for all to see that science was an essential key both to national defense and to ecoıomic prosperity in the modern state" (Thackray 1983, ›. 226).

With the war as a tremendous stimulus to R&D in all ectors, the federal government became a major supporter f research and has remained so ever since. Whereas ex-enditures from all sources for scientific research and de-elopment in the United States totaled about $345 million n 1940, the government alone was spending more than wice that amount near the end of the war ($720 million in 944). By the late 1950s, the total for all R&D expenditures n the United States reached $10 billion, of which the fed-ral government supplied two-thirds (Thackray 1983, pp. 27, 232).

The research universities were major benefactors of this ew federal largess. Federal support of academic research ontinued to grow after the 1950s and well into the 1960s see figure 2). It leveled off by the end of the decade and emained relatively flat until the late 1970s, when it in-reased again. Academic R&D expenditures from all ources totaled about $3.25 billion in 1981 ($6.6 billion in urrent dollars), with about two-thirds coming from federal ources. During the same period, industrial support of cademic R&D also continued to grow, but the proportion f support coming from industry fell dramatically from arlier levels as federal support increased, only rising again lightly as the flow of federal funds began to slow (see fig-re 3).

From the late 1960s to the present, the overall pattern of esearch and development in the United States has exhib-ed a fairly stable mix: basic research at about 13 percent, pplied research at about 22 percent, and development at bout 65 percent. In industry, however, basic research as a ercentage of total R&D declined significantly after 1967, vith less money being spent on relatively risky and rela-ively long-term projects, while in the academic sector, asic research as a percentage of all R&D continued to ncrease as a result of federal funding (Hewlett et al. 1982, . 576). Various estimates suggest that by the beginning of he 1980s, basic research represented about two-thirds of ll academic R&D, or about $4.4 billion in expenditures National Science Foundation 1982, p. 7).

Government funding of academic research after the war einforced the universities' orientation toward fundamental

FIGURE 2
SOURCES OF SUPPORT FOR ACADEMIC R&D: 1960–1981

"Industrial support as percentage of total academic R&D expenditures"

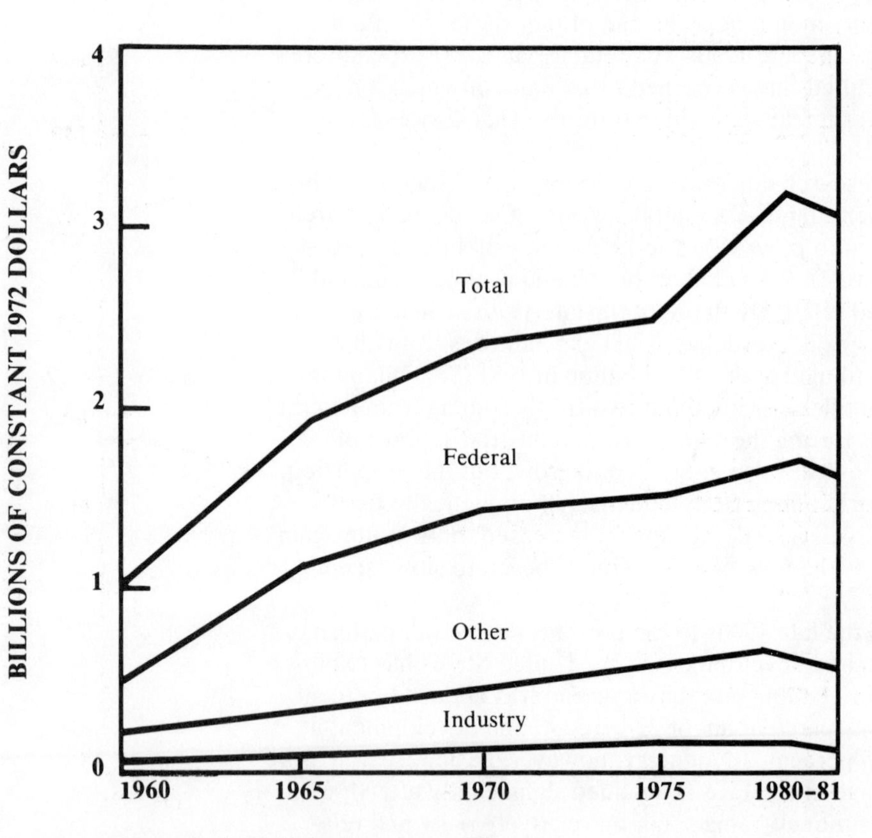

Source: National Science Foundation 1982, p. 7.

FIGURE 3
INDUSTRIAL SUPPORT OF ACADEMIC R&D: 1960–1981

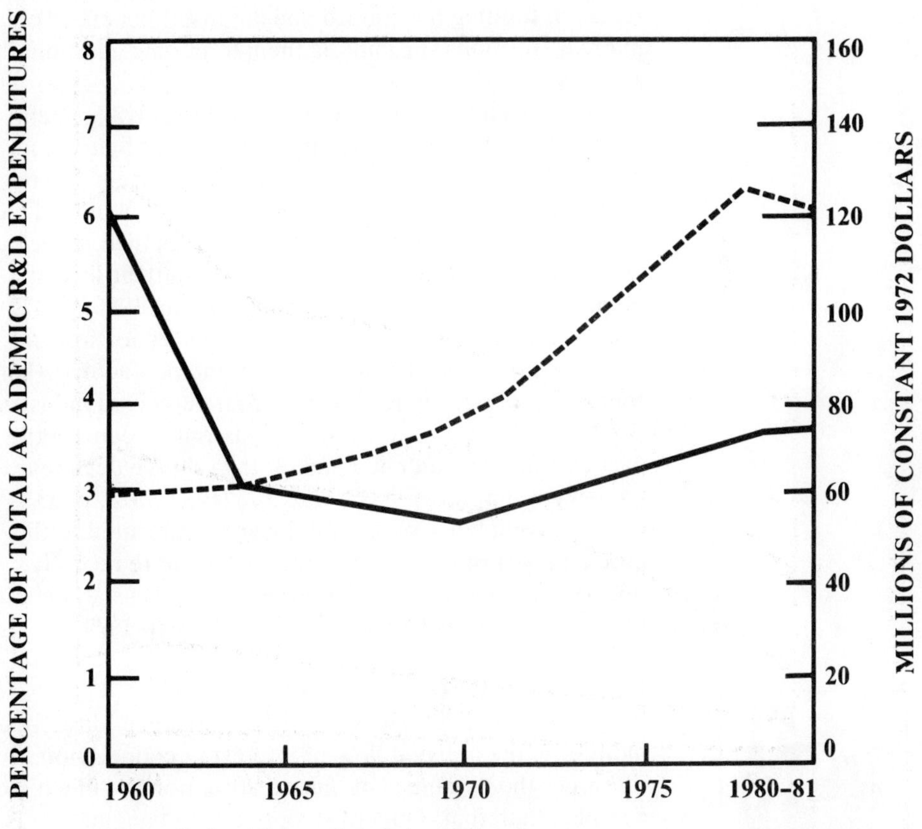

Source: National Science Foundation 1982, p. 5.

investigation and contributed to the expansion of graduate study and research programs, especially in the leading research universities. From 1971 to 1979, the top 100 institutions in R&D consistently received over 80 percent of all federal support for research awarded to higher education, creating a certain amount of vulnerability to alterations in the level of such support (McCoy, Krakower, and Makowski 1982, p. 325). The decline in the growth of federal research funding has indeed had an unsettling effect on many institutions, prompting them to pursue additional sources.

The 1950s and 1960s saw close working arrangements develop in aerospace and a number of other fields, and high-technology firms were spawned from academic research centers in such places as Boston and Stanford. As a general trend, however, relationships atrophied in the post-war period (Cooper 1979, p. 27), reaching their lowest point in the early 1970s (Prager and Omenn 1980, p. 379). This weakening of bonds between higher education and industry can be attributed to three principal factors: (1) the separation of academic research from perceived industrial needs; (2) decreased interest in industrial careers on the part of graduate students; and (3) the relative decline in industry's support of basic research (Baer 1977, p. 33). Greater availability of federal dollars contributed to this moving apart of the two sectors, just as more recently, uncertainty about its availability is regarded as an impetus for renewed research relationships (Shapero 1979).

Incentives for Research Linkages

Although financial factors clearly do influence, directly or indirectly, the ebb and flow of interest in cooperation in research, the dynamics of mutual attraction are much more complex than that. One must look at the changing needs and pressures of each sector, as well as general trends rooted in the changing nature of technology.

In the late 1970s, academic institutions were "seeking to rediscover their traditional ties to industry," partly because of a "real interest in the difficult issues relating to the . . . economy" and partly because of "more practical matters" (Baer 1977, p. 38). Some of these practical considerations included increasing competition for federal funds, governmental regulations reducing flexibility and freedom, generally worsened conditions of academic employment owing to enrollment and financial problems (Prager and Omenn 1980, p. 380), and deteriorating instrumentation in academic laboratories (National Governors Association 1983, p. 12). But increased interaction with industry may provide other benefits as well. Industry is a potential employer of graduates, a source of part-time faculty, a focus for continuing education programs, and a source of new ideas to students and faculty through exposure to current industrial problems (Prager and Omenn 1980, p. 380).

These other benefits were among those cited by college and university presidents in a recent survey of 180 institutions (Ferrari 1984). In fact, they were rated as considerably more important than the goal of attracting corporate funds. Still more important was the general goal of contributing to state and regional economic development, which more than three-fourths of the presidents surveyed rated as the single most important reason for academic involvement with industry (Ferrari 1984, p. 6).

Data based on interviews with directors of cooperative programs at 39 universities present a somewhat different picture (Peters and Fusfeld 1983). New sources of money was cited most often as a "motivating factor" at 41 percent of the institutions studied, followed by students' exposure to industry (at 36 percent of the institutions) and improved training of students (at 33 percent of the institutions).

Differences in methodology, sample size, and respondents' positions make it difficult to compare the results of these studies, though one might expect academic leaders, particularly in public institutions (90 percent of Ferrari's respondents) to formulate goals in broader terms of public interest than might be the case for academic research directors attempting to sustain or strengthen their own research programs. Further, the presidents surveyed by Ferrari tended to look primarily to state sources for support and, indeed, regarded its insufficiency as the most serious barrier to cooperation with industry (p. 37).

From industry's point of view, the potential benefits from research relationships with academic institutions include access to competent scientists, sources of potential research employees, stimulation of industrial scientists and engineers, and sources of ideas, knowledge, and technology for new products and processes (Prager and Omenn 1980, p. 380).

The usefulness of such lists of needs or potential benefits tends to be somewhat limited by virtue of their generality. Industry may look to higher education for the "advancement of scientific and engineering frontiers," for "well-trained graduates," and for "appropriate technical services" (White and Wallin 1974, p. 30), but under what conditions are such expectations likely to lead to concrete agreements for cooperation in research? Empirical studies of interests or motivating factors provide some insights, if not completely satisfactory answers. For example, a survey of 130 industrial research executives conducted several years ago found that the primary areas of interest were discussion of alternative solutions to specific problems, joint generation of new ideas for products and services, and transfer of new technology through communiques, theses, liaison, and consulting relationships (Libsch 1976, p. 30).

A logical inference might be that industry seeks to learn as much as it can from academic science, preferably on specific matters of the most immediate interest. A number of sources suggest, however, that corporate research executives are among the first to suggest continued emphasis in the universities on fundamental research (Cromie 1983; Libsch 1976; Peters and Fusfeld 1983). Moreover, surveys of industrial leaders reflect less reliance on academic expertise for innovations than might be expected (Peters and Fusfeld 1983; Sirbu et al. 1976). A possible explanation is the different concepts of "innovation" found in academia, where the term implies a breakthrough, a whole new idea or approach, and in industry, where it refers to later stages of the process or sometimes to the total process from concept through development and commercialization (Peters and Fusfeld 1983, p. 36).

Nevertheless, calls for better communication, for increased personal interaction, and for other arrangements to increase mutual understanding are legion in the literature. Interest in tapping academic expertise frequently prompts corporations to look for the most qualified people nationally, because excellence is often more important than proximity (Sirbu et al. 1976; Sorrows 1983) and because targeted support of leading researchers may be less expensive than maintaining similar competence in-house (Cromie 1983); Shapero 1979). This kind of selective support occurs most often when a gap exists in corporate knowledge of a particular research area or when, in the case of emerging fields like biotechnology, intense competition and the probability of rapid commercialization exist (Culliton 1982).

This combination of circumstances—increased competition and new needs for knowledge in science-based technology—is expected to affect many industries in the 1980s and beyond. "The problems and opportunities in technologically based industrial production are substantially different from those of the past":

- Product and process improvement in many fields are more complex, demanding an understanding of fundamental physical and biological phenomena.
- Incremental advances in narrow, technical areas are giving way to the use of a broad range of disciplines and analytical capabilities for problem solving.
- The diffusion of research capabilities makes it increasingly unlikely that any one company can retain an exclusive hold on progress in a particular area of technology (National Science Foundation 1982, pp. 16–17).

These factors are interactive, and we may see them "converging to create new configurations of academic and industrial research" (p. 16).

Research Relationships: Patterns and Models

The kinds of research interactions that develop depend on organizational characteristics and environmental conditions. For *industry,* those characteristics include size, structure, profitability, nature of business, and progressiveness of the research program; for *academic institutions,* they include the type of institution, size, financial health, stature of scientific and engineering programs, and the orientation of research programs. The *external factors* that play a role include geographic proximity, location of alumni in key positions, and the migration of faculty to industry and of industry personnel to academia (Prager and Omenn 1980, p. 381).

Academic/industrial research ties tend to be concentrated at the leading universities and among relatively few, mostly large corporations. Only about 200 institutions have research expenditures substantial enough to be considered "major research universities"; 100 of them receive over 80 percent of all the federal funding of academic R&D (Fusfeld 1983, pp. 12–13). On the industrial side, the distribution of R&D is also skewed, with the top 10 companies in R&D accounting for 33 percent of the expenditures. The major R&D industries are chiefly in the chemical, pharmaceutical, electronic, computer, fuel, aerospace, automotive, and petroleum fields. A disproportionate amount of the support provided to universities by these industries goes to professionally oriented schools and departments, especially in engineering, medicine, and agriculture (Peters and Fusfeld 1983, p. 49).

Many new cooperative arrangements for research have been established as a result of faculty initiative and often as a consequence of prior consulting connections (Peters and Fusfeld 1983, pp. 18–21). A multistage pattern of relationships thus emerges:

- *Company wants technical/scientific advice (general or specific) and seeks out professor;*
- *Professor/consultant sees opportunities for research and initiates research relationship;*
- *Company tracks and (maybe) utilizes the research, and makes employment offers to the bright graduate students and postdoctorates working on the project;*
- *The cycle is repeated in future years* (National Science Foundation 1982, p. 29).

Research relationships can be characterized in a number of ways. The type of exchange that occurs may focus on information, on people, or on resources (National Science Foundation 1982, pp. 20–21). The duration of the interaction may be limited or continuing (Brodsky, Kaufman, and Tooker 1980, p. 1). Or the arrangements may relate to various policy objectives (Baer 1977). The primary mechanisms for such cooperative relationships include research centers and institutes, industry-sponsored contract research, special university/industry research agreements, personnel exchange programs, research consortia, and cooperative research centers.[1] In the literature, descriptions of the major vehicles for cooperation in research are becoming increasingly standardized, though definitions do not always agree.

Research centers and institutes

Over 5,000 research centers are operating in colleges and universities, many of them for the application of academic research to the problems of industry or other sectors of society (Shapero 1979, p. 30). Three reasons have been suggested for the proliferation of these units: (1) to permit academic pursuits that do not fit into departments offering conventional subject matter; (2) to accommodate the interests of academic innovators who prefer "an end run" to launch programs rather than working through normal structures; and (3) to respond to external opportunities more quickly than possible through departmental mechanisms (Wilson 1979, p. 86). "The institute . . . has been as much the vehicle of innovation in recent years as the department has been the vault of tradition" (Kerr 1972, p. 102).

At many institutions, a major share of industrial support is channeled through research centers.

Variously named centers, bureaus, institutes, or laboratories, these organizational entities serve as a focal point for special research interests and activities. They may be located within a department or a school, or they may be affiliated with the institution but organized as a separate unit. External funding is common, though sources vary and often represent a mix of government, foundation, and industrial funding as well as support from the institution. At many institutions, a major share of industrial support is channeled through research centers (Finkbeiner 1969, pp. 1–9).

Research centers offer several advantages for cooperation with industry (Peters and Fusfeld 1983, pp. 32, 82, 106–7). The function of focusing research endeavors can be aided through a central facility, special equipment, or simply by bringing intellectual coherence to some general area of research. Moreover, centers can concentrate manpower

[1]Research parks are discussed in the next section as a form of technology transfer. Though often intended, in part, to stimulate research relationships, they are a subspecies of industrial park and typically serve a number of purposes related to business and economic development. Faculty consulting is also covered in the next section.

in a multidisciplinary setting for various time periods, and projects can be terminated with minimal disruption. In some cases, foundations or separate corporations are established to act as buffers or transitional structures, making it easier to interact with industry and, in the case of public institutions, minimizing the constraints of state government regulations. Control typically resides within the academic institution, with industrial representatives serving on advisory boards.

Research centers may focus on regional needs or on the needs of specific industries. For example, programs supported by the paper industry exist at Lawrence College, Wisconsin, at the State University of New York's College of Environmental Science and Forestry, and at North Carolina University's Pulp and Paper Research Institute (Brodsky, Kaufman, and Tooker 1980, pp. 19–20). These centers conduct basic research, disseminate information to industry, train graduate students, offer seminars, and test new products. As these examples from the paper industry suggest, academic research centers may be the focal point for a variety of types of academic/industrial interaction. They may, for example, take on multiple sponsors as part of a research consortium or cooperative research effort.

As organizational entities, however, research centers are not without their problems. Their staffs may or may not have regular academic status, and their dependence on external funding requires that center directors spend considerable effort in promotion (Wilson 1979, p. 86). They may also lose touch with the institution's discipline-based departments, thereby diminishing rather than strengthening ties between the academic mainstream and the industrial community (Shapero 1979, pp. 30–31). Nonetheless, research centers appear to be an important component of higher education's relations with industry, and evidence suggests that many institutions are planning to expand such programs in the near future (Ferrari 1984, pp. 8–9).

Industry-sponsored contract research

A second type of interaction is contract research—investigation conducted under contract between a company and a college or university. The work agreement is usually with one or more individual professors but is legally contracted with the institution itself. Overhead must e calculated, institutional policies adhered to, and approval received from the appropriate academic units. Over 0 percent of industrially supported research in universities s funded through the mechanism of contract research (Peers and Fusfeld 1983, p. 71).

Corporations look to academic institutions for research under contract when a gap exists in corporate knowledge of a specific area, when an institution owns sophisticated instrumentation not otherwise available to the firm, or when evaluation is required of materials the company has developed (Brodsky, Kaufman, and Tooker 1980, p. 31). Typically involving small amounts of money—often less than $50,000—such arrangements are also usually short term and negotiated project by project or year by year. A high degree of monitoring may occur because "the industry [liaison] must continually justify and resell the project to his superiors at every budget hearing" (Roy 1972, p. 956). The mission-oriented nature of such industrial research contracts means that corporate sponsors are highly concerned about proprietary rights. As a result, publication of findings and patent policies may constitute serious issues (see "Academic/Industrial Cooperation in Perspective").

Despite these drawbacks and difficulties, most writers regard contract research as a desirable form of research relationship, one that generates vigorous interactions between academic and industrial personnel. It meets an important "market test" because it depends on the perceived value to the sponsoring firm of the resulting specific research findings and information (Baer 1980, p. 10). Further, such contract research may lead to larger-scale, longer-term research agreements or be extended to multiple-sponsor programs, as in the case of research consortia (Baer 1980).

Special university/industry research agreements

No area of collegiate/corporate relations has received as much attention of late as the handful of special research agreements signed in the last few years by major corporations and leading universities.

Academic/industrial cooperation became a media event when the Monsanto Company announced in February 1975 that it had agreed to provide up to $23 million over a 12-

year period for biological and medical research at Harvard University. This type of arrangement—a long-term, high-level commitment by industry to support basic science in return for closer links to university research and some proprietary advantages—is in contrast to the limited, short-term arrangements characterizing most contracted research (Smith and Karlesky 1977, pp. 66–67). This multi-million dollar example of academic/industrial cooperation was built on a foundation of personal interactions spanning 16 years (Prager and Omenn 1980, p. 382). The decision to support research on the biochemistry and biology of organogenesis grew out of Monsanto's long-range planning and management interest in its commercial possibilities. Monsanto funds are used at the discretion of the university to support research within the purview of a charter agreement in several Harvard departments. The respective roles of the university and the corporation in this venture are described as follows:

- *Harvard provides the conceptual scientific framework, identifies capable scientists, provides training, and controls relevant research at both Harvard and Monsanto during the research phase.*
- *Monsanto helps identify research needs, provides critical starting materials beyond the ability of Harvard to produce or buy, provides unusual and exotic analytical capabilities, and controls the development phase, providing expertise in technological innovation, development, and marketing* (Prager and Omenn 1980, p. 382).

Even more spectacular, in terms of level of funding, is the recent agreement between Massachusetts General Hospital, the teaching hospital affiliated with the Harvard Medical School, and Hoechst, A. G., a West German chemical company. This 10-year, $70 million program underwrites the costs of basic research in molecular biology in exchange for an exclusive licensing option and access for Hoechst personnel to the ongoing laboratory work (Bouton 1983, p. 63).

Additional examples of special agreements are cited in the literature:

- **Monsanto:** $23.5 million to Washington University for medical uses of proteins and peptides and $4 million over five years to Rockefeller University for research on plant genes;
- **W. R. Grace and Company:** up to $8.5 million over 10 years to MIT for combustion research;
- **E. I. DuPont de Nemours and Company:** $6 million over five years to Harvard Medical School for a new Department of Genetics;
- **Bristol-Myers Company:** $3 million to Yale University for the production of anticancer drugs;
- **Celanese Corporation:** $1.1 million to Yale University for enzyme studies;
- **Engenics Corporation:** $1 million each to Stanford University and the University of California–Berkeley over four years for research on chemical processes in genetically engineered microorganisms (Bouton 1983; Culliton 1982).

However exciting, such ventures are unfortunately quite rare, "both in the intensity of resources committed and in the timeframe of the relationship" (Brodsky, Kaufman, and Tooker 1980, p. 32). They are also highly concentrated in a few emerging fields, such as biotechnology, and may represent a temporary mechanism by which some industries hope to catch up and build their own research capabilities. Even more important as a cautionary note is the observation that industry neither has the resources nor sees it as its responsibility to replace federal support of fundamental science for its own sake (Culliton 1982, p. 962). Still, as significant experiments in academic/industrial cooperation, these programs merit careful examination in terms of the new patterns of interaction they are evolving and in terms of the way they are resolving problems that have long confronted research relationships between the sectors.

Personnel exchange programs

Personal interaction between representatives of industry and representatives of academia occurs to various degrees in all of the cooperative research structures described here, but not necessarily as an objective in its own right. The assumed value of such interchange among research personnel in particular has prompted a number of companies and institutions to establish special programs to promote the exchange of personnel.

The assumption that personal interaction is beneficial has some basis in the empirical literature. In a study of research productivity among scientists in industry, government, and universities, outstanding performance was highly correlated with both the level of communication with colleagues and the diversity of work (Pelz and Andrews 1966, pp. 51–52, 65). The leading scientists tended to have extensive contact with other researchers and tended to participate in two or more R&D functions rather than just one (applied as well as basic research, for example). Analysis of patterns of technology transfer has disclosed that "inventors" and "exploiters" are often different kinds of people and that mechanisms to link such persons are likely to yield greater payoff in terms of the commercialization of inventions (Roberts and Peters 1981, pp. 122–24). As noted earlier, the origins of academic/industrial programs in prior consulting relationships has also been documented (Peters and Fusfeld 1983, p. 40). Whatever the evidential basis may be, the assertion that direct contact and communication between academic and industrial researchers are crucial for productive relationships is one of the most widely accepted generalizations in all of the literature (for example, Baer 1977, p. 49; National Science Foundation 1982, p. 23; Sorrows 1983, p. 6).

Academic associations, foundations, and corporations have sponsored formal arrangements enabling academic and industrial scientists to spend periods of time in laboratories operated by their opposite numbers. The American Physical Society initiated the Visiting Physicists Program in 1973 for short visits of one or two days, with exchange occurring in both directions, and some universities—the California Institute of Technology, for example—have similar programs. The Ford Foundation in 1963 launched the Residencies in Engineering Practice Program, allowing academics to work as long as 15 months in industrial settings. Marathon Oil, DuPont, General Electric, and IBM are among the corporate sponsors of exchange programs featuring summer employment, part-time work through a year, or visits of from one week to several months

(Brodsky, Kaufman, and Tooker 1980, pp. 68–69; Peters and Fusfeld 1983, p. 87).

While lengthy sojourns are uncommon because of scheduling, family responsibilities, and career obligations, many researchers participate in limited visits, find them attractive, and favor their expansion. About 35 percent of the 180 institutions in one study have some kind of personnel exchange program, and 70 percent of them anticipate having such a program by 1988 (Ferrari 1984, p. 11).

One variation on this mode of collaboration is to form teams of corporate and collegiate researchers to survey the needs and opportunities for new technology in selected industries (Stever 1972, p. 26) or to convene joint meetings between corporate research managers and selected academic departments to acquaint faculty with current industrial needs (White 1973). Faculty members are often unaware of what is occurring in neighboring companies and might "catch some of the excitement and be stimulated to do research that is both on the frontier and very relevant" (White 1973, p. 14).

Research consortia

The research consortium is listed as a separate model of academic/industrial cooperation in most sources, but different writers describe it differently and the mechanism tends to merge conceptually with other modes of cooperation—an inevitable occurrence as one moves from simple bilateral agreements or from single-form exchange programs to multiorganizational and multipurpose arrangements.

In fact, the clearest definition of the research consortium (Prager and Omenn 1980, p. 381) is somewhat misleading. Although defined as a "single university–multiple companies" model, research consortia can also involve multiple institutions and multiple companies, or multiple institutions and a collective industrial research association. The Processing Research Institute (PRI) at Carnegie-Mellon University illustrates the first type—single university, multiple companies—with more than 25 companies funding research projects of interest to the processing industry at Carnegie-Mellon. The Michigan Energy and Resource Research Association (MERRA) illustrates the second type—a consortium of several universities (in Michigan) and several large companies (including Dow Chemical and Detroit Edison). The third type is illustrated by programs sponsored at several universities by the American Petroleum Institute and by the American Iron and Steel Institute (Prager and Omenn 1980, p. 381; Roy 1972, p. 956; Smith and Karlesky 1977, pp. 67–69).

Both the PRI consortium at Carnegie-Mellon and MERRA also involve government as another participating entity. The National Science Foundation funds part of PRI, and the federal Energy Research and Development Administration and state government are involved in funding MERRA. Indeed, efforts to obtain federal and/or state funding are often the impetus for organizing consortia (Smith and Karlesky 1977, p. 69).

In addition to involving multiple organizations (in various combinations) in cross-sector cooperation, consortia typically focus on university-based research of generic interest to an industry, often assess membership fees to participating companies, and share research results among participants (Prager and Omenn 1980, p. 381).

Research consortia may recommend themselves to corporations wishing to benefit from generic research at universities but facing financial constraints. Companies are obliged to provide profits to investors in the near term and "simply cannot afford to fund, on an individual basis, research so basic as to offer, one day, the creation of a new industry" (Kiley 1983, p. 65). Although the companies participating in PRI and MERRA are among the nation's largest corporations in annual sales, firms other than industrial giants can participate in university research through consortia. The joint use of laboratories or joint purchase of expensive equipment, for example, could be undertaken by a consortium, though such examples are not common at present (Brodsky, Kaufman, and Tooker 1980, pp. 23–24).

Although the level of interaction may be less intensive than in other forms of research cooperation (Roy 1972, p. 956), research consortia sometimes exhibit a substantial amount of mutual involvement in planning and implementing research activities. This situation appears to be true, for instance, of the Gulf Universities Research Consortium, in which academic and industrial personnel at both the research and executive levels participate as directors, as policy council members, and as members of advisory panels (Sharp and Gumnick 1980, pp. 19–20). In such cases, no clear dividing line differentiates the research consortium from what is described below as the university/industry cooperative research center. What *is* clear is "how hard it is to generalize about and categorize consortia" (Patterson 1974, p. 23).

Cooperative research centers
In an attempt to categorize research relationships found in 39 universities and 56 companies, Peters and Fusfeld (1983) conclude that intersector arrangements constitute "a spectrum of cooperation," with contracted research projects at one extreme and "intimate collaboration in research design and management" at the other (p. 16). Cooperative research centers, located toward the upper end of the spectrum, have not only support from several companies (as do consortia) but also an advisory structure for industrial input and an industrial affiliates program for dissemination. Cooperative centers differ from consortia in that participating companies play "an active role in making policy, planning research, and overseeing the implementation and evaluation of research" (General Accounting Office 1983, p. 21).

For present purposes, cooperative research centers can be described as multiorganizational, multipurpose, and jointly planned and/or managed endeavors focusing on research and related activities of mutual interest. They exhibit the breadth of participation characteristic of consortia, the many-sided interaction found in special research agreements, and, in at least some cases, a high level of mutuality in governance as well.

Quite recently a number of state governments have sought to encourage cooperative research centers under the rubric of advanced technology application programs. Such federal agencies as the Department of Defense, the Department of Energy, the National Aeronautics and Space Administration, and the National Science Foundation (NSF) have a longer history of fostering cooperative research endeavors between academic and industrial organizations. Especially well known is NSF's University-Industry Cooperative Research Centers Program, in which funds must be matched by corporations for cooperative programs designed to become self-sustaining over a period of five years (Baer 1980, pp. 12–13; NSF 1982, p. 15).

Cromie (1983) describes seven case studies of university-industry cooperative research centers in the microelectronics industry: Stanford Center for Integrated Systems; Caltech Silicon Structures Project; Microelectronics Innovation and Computer Research Operation (Berkeley); MIT Microsystems Program; Microelectronics Center of North Carolina (Research Triangle Institute); Microelectronics and Information Sciences (University of Minnesota); and National Research and Resource Facility for Submicron Structures (Cornell) (pp. 235–54). A comparison of key aspects of these programs is instructive. All seven have been established in recent years (the center at Cornell, launched in 1977, is the oldest) but build upon well-established ties between participating universities and industrial sponsors. Four of them involve more than one university, and all have numerous industrial participants, mostly large, high-technology corporations with a strong orientation toward R&D.

These programs focus resources on leading scientific and engineering centers where advanced research and training is taking place that is of interest to the microelectronics industry. The activities at these centers cover a range of academic/industrial interactions, reflecting their multipurpose aspect:

- fundamental research on critical problems, such as the design of very large scale integrated systems;
- visiting scientist programs, in which industrial researchers spend up to a year in university laboratories;
- advanced training of graduate students (and, in North Carolina, training of technicians in high demand areas of technology as well);
- continuing education programs and technical seminars to update practitioners;
- industrial affiliate programs providing a window on technology through exchange visits, research reports, conferences, and symposia.

Sources of funding for these cooperative projects represent a mix of governmental and industrial support. In all cases, ongoing federal research contracts to the universities provide an essential base of faculty and student support, and in a few cases special grants from NSF or other federal agencies played a key role in launching the project. State support has also been important in some cases, as in North Carolina ($24 million in 1981).

Industrial support typically occurs in the form of special pledges or membership fees, ranging from $100,000 per year at Caltech to $250,000 per year at Stanford, MIT, and in North Carolina. Participants in California's MICRO program were asked to match state funding (about $1 million in 1981) through donations of cash and equipment. Minnesota's project was started with commitments of $1 million from each of four leading corporations in the industry. In a different approach, the Cornell project receives the bulk of its support from NSF, while companies pay modest fees for use of laboratory facilities.

These centers exhibit diverse patterns of governance. The programs at Caltech and MIT had no industrial advisory committees (as of 1982). The Stanford and University of California programs have advisory structures for industry, but their review or policy committees are composed of academics. The center at Cornell and the program at Minnesota have industrialists as well as academics on policy boards. The Microelectronics Center of North Carolina appears to be the only case where governance is fully shared between universities and corporations, with both groups represented on an advisory board and on the board of directors.

It is also interesting to note the expectations of industrial representatives regarding these programs. A vice president at Honeywell explained his company's participation in the Microelectronics and Information Sciences (MEIS) program at Minnesota:

> *"The principal reason we contribute to MEIS is to assure ourselves of an adequate supply of well-trained people. We expect to achieve this. We also hope for, but do not count on, a synergy between Honeywell's research program and the program at the University. Finally, we hope to contribute to the health of the whole industry by supporting long-range, fundamental research in high risk areas"* (Cromie 1983, p. 249).

According to Cromie:

- There is no single, right approach to cooperative ventures because needs and circumstances vary.
- Federal support is an essential underpinning of the universities' basic capabilities, without which such university/industry programs could not survive.
- Industrial participants favor team approaches and exchanges between academic and industrial personnel as opposed to situations where professors conduct research in a vacuum.
- The "sociological" aspect of cooperative programs is likely to be more difficult than the technical (Cromie 1983, pp. 251–52).

The success of cooperative research centers requires strong research capabilities within the institutions, an interest in and commitment to working on large-scale programs of importance to industry, and strong leadership in the management of the centers (Tornatzky et al. 1982, pp. 10–13).

Expanding the Agenda of Cooperation

There is nothing mechanical or easy about developing research links between academia and industry:

> *The process of establishing university-industry interactions is not linear; it is circular, iterative, and sometimes discontinuous. It is . . . an exercise in mutuality where understanding is more important than contracting; where personal contacts outweigh administrative mechanisms; and where ostensible purposes shelter undefined and even more valuable priorities* (NSF 1982, p. 23).

And the research interactions that may emerge through these models of cooperation are likely to be conditioned by several factors: the context in which they evolve, the nature of the respective institutions and industries, the capabilities, constraints, and needs of each sector, and the mismatches between them (Shapero 1979, pp. 3–5).

Some of the limitations associated with each of these models of cooperation for research have already been noted. The academic research center, as a bridge between external groups and the traditional departmental structure, may not be fully integrated into the latter. The research contract raises issues concerning proprietary rights. The special, long-term bilateral agreement, however impressive, occurs seldom and in limited areas of research. The exchange of personnel is limited by scheduling and by personal or career constraints. Consortia may sacrifice depth of interaction for breadth of participation.

Moreover, these difficulties have a cumulative aspect for the models of relatively greater complexity. The cooperative research center in particular must deal with problems attendant to all of the other models: Its home base is usually an academic institute, one of its modes of operation is likely to be contract research, its multiple membership may make interaction with individual companies difficult, and so on. While the probability of greater benefits exists, it is clear that increasing complexity brings increasing challenges in terms of institutional commitment and interorganizational leadership.

A dynamic that has been observed in other kinds of interorganizational relationships may be at work here: As academia and industry draw closer together, additional avenues of mutual interaction may become apparent. A university can present "a broader interface for cooperation with industry if it so desires," and institutions can "package more interdisciplinary programs" and "present a wider array of services" (Peters and Fusfeld 1983, p. 126).

Expanding the agenda of cooperation may reflect not only a response to additional opportunities but also the recognition of additional needs. Understanding the complete cycle of innovation (and its different meanings to academics and to industrialists) has prompted many to call attention to the need for the effective transfer of technology. The concern for technology transfer, in fact, constitutes another major dimension of academic/industrial relations and requires separate treatment.

COOPERATION IN TECHNOLOGY TRANSFER

The complexity of the innovative process reflects the economic difference between the creation of scientific knowledge and the application of this knowledge to the advancement of economic welfare. . . . The intervening processes are often long, complex, and costly. Much of the cost and time are associated with the stages beyond the generation of the basic technology itself, specifically, with the production and marketing of new products made possible by new technology (National Academy of Sciences 1978, pp. 12–13).

An important caveat accompanies the current interest on the part of economic planners, community growth associations, government officials, and business leaders in the fruits of academic research. From the viewpoint of economic development, academia is an important resource only insofar as its contributions are tied to economic growth. The justification for increased R&D expenditures, from this viewpoint, lies in the actual implementation of research results (Hewlett et al. 1982, p. 581). Even support for academic/industrial partnerships in advanced technology may be provided only when "a tangible relationship to product or process commercialization and hence job creation can be demonstrated" (Holtzman 1983, p. 9).

From [one] viewpoint . . . academia is an important resource only insofar as its contributions are tied to economic growth.

Economic development analysts generally agree that while advances in scientific and technological knowledge are indispensable for spurring innovation, this activity is part of a larger process by which new ideas are reduced to practice and introduced into the market. The desire to ensure that the fruits of research are used for technological development thus leads to an interest in ways in which the academic/industrial connection might foster the transfer of technology.

This chapter discusses the transfer of technology in relation to the stages and sources of technological change and the factors that influence it, reviews patterns of innovation in the industrial corporation and in the small, entrepreneurial firm, and outlines the formal mechanisms by which technology transfer is facilitated.

Technology Transfer

Outside of agriculture, government exhibited little interest in transferring technology until the mid-1960s. As federal programs expanded and competition for budget allocations increased, agency officials sought evidence of the use of research, and a number of programs "to promote the spread of technology" were launched (Tornatzky et al. 1983, p. 161). Uncertainty concerning the most effective mechanisms remains, however, and agricultural extension is still regarded as the classic model (pp. 162–69). The success of the agricultural model lies in the way in which "extension has linked the research and educational facilities of the agricultural colleges with the farmer who is the user of the technology produced" (General Accounting Office 1983, p. 35).

The mechanisms of technology transfer are "programs structured with a view to capitalizing on university research or integrating technological results of university research into private sector programs or commercial products" (Peters and Fusfeld 1983, p. 98). Such mechanisms may address specific research problems in industry, promote brokerage and licensing, or provide technical assistance to existing or new companies. Putting somewhat greater emphasis on the role of entrepreneurs in bringing technology to the marketplace, Levy (1977) prefers the term "commercialization," arguing that it is a broader concept than technology transfer. When such individuals form their own businesses, the result is "significant technology transfer as well as impressive commercial and economic impact" (Roberts 1968, p. 258). Others equate commercialization with the later stages of the change process as a whole.

The nature of what was transferred by means of entrepreneurial exodus varied considerably in Roberts's sample, however. In some cases ("direct transfer"), the technology itself was the essential element, without which the new company could not have been started. In other cases, a source technology was augmented by others ("partial transfer"), or no traceable transfer occurred at all ("no transfer") (Roberts 1968, pp. 260–61). Nevertheless, one of the major correlates of success among spin-off companies is a high level of technology transfer: The highest performers tend to be those substantially grounded in a source technology (p. 263).

It is often far from clear whose responsibility it is to ensure that transfer occurs (Prager 1983, p. 10). Not everyone, in fact, agrees that commercialization is an appropriate aim, at least as a public policy objective. The pressures for financial payoff from federal R&D are "misdirected concerns for commercialization," and governmental policies should not mirror the short-term orientation of industry (Rettig 1982, p. 400).

The Process of Technological Change

The phenomenon of innovation has been examined from many perspectives, using different conceptual and methodological approaches. The findings of these examinations are not cumulative, however, because interdisciplinary work on synthesizing and integrating findings from various fields has been insufficient (Tornatzky et al. 1983, p. 47). Definitive work still lies ahead:

> *The relative importance to innovation of individual behavior, group dynamics, organizational context, and economic/societal factors, and how these influence and condition each other, remains a major question for future research* (Tornatzky et al. 1983, p. 219).

If one considers a single technological innovation, occurring in a particular time and place, the variables that directly or indirectly affect it seem almost endless. First is the technology itself and its sources and antecedents; then add to it the individual or individuals involved, with their knowledge, skills, and other characteristics, the organizational or interorganizational setting, the environment within the broader community, state and federal policy factors, and the conditions in the international market. It is not surprising, therefore, that the relationship between R&D and technological innovation and the precise effects are not well understood (see, for example, Abernathy and Rosenbloom 1982, p. 418; Committee for Economic Development 1980, p. 63; National Academy of Sciences 1978, p. 18; Tornatzky et al. 1983, p. 219).

Nonetheless, a number of major themes are apparent in the literature: the stages of innovation, the sources of technological change, and the factors thought to facilitate technological development in communities. Beyond these topics are considerations of innovation in existing industries and among new entrepreneurial firms, both of which shape academic/industrial cooperation in technology transfer in important ways.

Stages of technological change

Students of innovation generally define it in terms of "all of the activities engaged in . . . when translating an idea or concept into the economy" (Lamont 1971, p. 43). In industrial use, however, the term "innovation" often refers to one particular stage in the chain of a longer process, and it is *not* the conceptual stage. The conceptual stages are described as "invention," and the term "innovation" refers to the stage in which an idea is introduced into the marketplace by setting up the first facilities for manufacture and marketing (Committee for Economic Development 1980, pp. 13–14; Holloman 1974, pp. 6–9). The final stage is "diffusion," or replicating in other plants the products and processes that have proven successful.

Time, cost, and risk increase as one moves through these stages, accruing primarily to the industry involved (Committee for Economic Development 1980, p. 17). Although change occasionally occurs rapidly through revolutionary breakthroughs in the laboratory, the general pattern is evolutionary, a lengthy process of small, incremental changes in design or practice (Holloman 1974, p. 7; National Academy of Sciences 1978, pp. 12–13). The development of penicillin illustrates this process. Alexander Fleming discovered the antibacterial characteristics of the mold, *Penicillium notatum,* in 1928. But penicillin as a substance was not isolated until 1938, after a decade of work by many people. Even after isolation, tens of millions of dollars and hundreds of man-years had to be invested before a clinically useful drug was possible and large-scale production feasible. Following 1944, when it was introduced, improvements continued over the next 20 or so years (National Academy of Sciences 1978, p. 13).

Sources of change

It is evident from this description of developmental stages that, in addition to new concepts themselves, many factors enter into the process. Because the flow from research to new products or processes is not automatic, much discussion has been given to the question of "technology push" versus "market pull." Market need is as important a determinant of change as is R&D, though research provides the basic tools or capabilities, and the formula for technological development can be represented schematically as follows:

> CAPABILITY + NEED → INVENTION → TECHNICAL DEMONSTRATION OF FEASIBILITY → INNOVATION → DIFFUSION (Holloman 1974, p. 7).

Going even farther in emphasizing the influence of market factors, Healey (1978) upholds the "wet noodle rule of innovation"—that "the pull of the market need is more effective for successful innovation than the push of technology" (p. 16).

A more balanced view is that scientific and technological advances are a necessary but not sufficient cause of change.

> *It is nearly impossible to predict whether a particular basic research project will lead to successful innovation. But it is clear that socially valuable innovations flow from high-quality basic research. Postwar innovations in atomic energy, computers, chemicals, pharmaceuticals, and other areas would have been impossible without a foundation of basic research results* (Committee for Economic Development 1980, p. 63).

In the current national climate of concern for productivity and economic growth, however, many feel that it is no longer satisfactory to leave the possibility of utilization to chance. To conceive of the academic role strictly in terms of fundamental inquiry means that academia is not really part of the innovative process but merely part of the environment of user organizations (Tornatzky et al. 1983, p. 170). Or, put another way:

> *The assumption . . . that a number of unutilized ideas exist in universities is not per se useful until we learn how an interested commercial organization can take advantage of or even become aware of those ideas in which it might have interest* (Roberts and Peters 1981, p. 123).

Two of the most important findings in the literature are that integrating R&D with marketing is critical for innovation and that face-to-face communication has a strong, positive effect on dissemination (Tornatzky et al. 1983, pp. 156–60). Those findings have an important implication for technology transfer:

> *The more a technology transfer system encourages direct communication between source and user in the choice of the knowledge to be transferred, the greater will be its success as seen by both sides* (Tornatzky et al. 1983, p. 220).

These findings underscore the importance of cooperative research models that make interaction between academics and industrial personnel closer; further, they suggest that specific measures should be taken to foster mutual understanding of developments and problems in both sectors.

Community factors in development

Another factor in technological change relates to the conditions existing in a community, state, or region. As noted earlier, numerous state initiatives focus on creating a climate conducive to development; the same is true in certain communities. While many studies focused on community initiatives suffer from a narrow conception of development, one important exception is the study of 28 "technology-oriented complexes" (TOCs), conducted by Sirbu et al. at MIT in 1976, which disclosed four distinct patterns by which these technological complexes evolve:

1. TOCs develop principally as a product of spin-offs and locally initiated companies (for example, Boston's Route 128, Palo Alto, Ann Arbor);
2. TOCs develop in a park site with public and private R&D facilities, excluding or limiting manufacturing (for example, Research Triangle Park, North Carolina; Sheridan Research Park, Ontario);
3. TOCs develop by attracting the manufacturing facilities of high-technology companies (for example, the Edinburgh-Glasgow belt in Scotland; Phoenix, Arizona);

4. TOCs result from heavy government spending at a development facility (for example, Huntsville, Alabama; Houston, Texas) (pp. 20–26).

These communities have several factors in common: (1) a high concentration of scientists and engineers; (2) diverse organizations and institutions that employ professional people; (3) some degree of interaction among these organizations; and (4) a self-perception or self-image of their area as technologically oriented (pp. 8–9). They also enjoy similar benefits: high salaries, attractiveness to professionals, growing industries, and increased employment opportunities (p. 65).

None of these patterns of development are superior to the others: The best approach is the one that fits the aims, needs, and resources of a particular community. Certain factors are important for the development of such TOCs, however: positive local initiatives and attitudes, governmental financial incentives, and the presence of academic institutions. Universities seem to be more important as "a cultural amenity for attracting new engineers and scientists" than for their research contributions (p. 41). Although, in Sirbu et al.'s study, attending courses and colloquia was common among professional employees, research ties with local universities were strongly developed in fewer than half the cases.

"The most significant interaction between industry and the educational system . . . took place on the junior college or technical school level" (Sirbu et al. 1976, p. 45). Two-year colleges, especially, supplied technical and laboratory assistants. While universities often held "an unfavorable attitude toward close industry relationships" (p. 56), community and technical colleges tended to be "quite willing to modify their courses and curricula to suit the needs of local industry" (p. 46).

Universities could play a much greater role in liaisons with industry, and benefit far more from its presence, by deliberate steps "to begin breaking down the barriers" (p. 72). With larger companies, such steps might include joint meetings between research directors and faculty; with new ventures, they might include encouraging spin-offs and assistance to entrepreneurs. Education and service links and research connections should be pursued, with different roles for different kinds of institutions.

Innovation in Industrial Corporations

Although much of the current literature concentrates on the small, entrepreneurial enterprise, earlier studies tended to make the established industrial corporation their subject. As academic/industrial research links most often relate to large corporations, the process of innovation, as it occurs in these organizations, is an important part of the context for technology transfer.

A recent blue ribbon commission in Great Britain, the Committee of Inquiry into the Engineering Profession, argues vigorously that innovation among existing industries is essential for the prosperity of advanced industrial nations:

> *Continuous innovation has of necessity become a way of life for successful manufacturing companies to cope with the inherent obsolescence of products and production methods in a continually changing technological and market environment* (Finniston 1979, p. 26).

The kinds of innovations required to meet or anticipate market changes may be grouped under three headings:

1. *Incremental improvements to existing products, production methods, and processes;*
2. *Diversification, using existing expertise and capabilities in different product markets; and*
3. *Radical departures from previous activities, based on the introduction of products or processes embodying novel applications of technology* (Finniston 1979, p. 26).

The appropriate innovative strategy will vary with the circumstances of particular companies and markets.

In this country, some of the most definitive research on industrial innovation has been done by Edwin Mansfield. According to Mansfield (1968), the rate of technological change in an industry depends on many factors:

- The amount of resources expended on R&D to improve an industry's technology;

- The resources devoted by other industries to improve the capital goods or other inputs an industry uses;
- The industry's market structure;
- The legal arrangements pertaining to its operations;
- The attitudes toward technological change of management, workers, and the public;
- The way internal R&D activities are organized and managed;
- The way the scientific and technological activities of relevant government agencies are organized and managed; and
- The amount and character of R&D carried out in universities and in other countries (pp. 4–6).

mong these factors, Mansfield emphasized expenditures n industrial R&D as the most critical, and he devoted a ood deal of his analysis to decisions affecting such expenditures.

In the firms Mansfield studied, the amount spent on :&D, whether in total or for specific projects, depended hiefly on estimates of profitability. Other variables inluded preference for safe projects over risky ones, an nterest in satisfying scientific as well as commercial objecves, and political factors such as executive pressure or roject managers' advocacy (p. 63). Owing to difficulties in stimating costs and the time necessary for completion, ecision making about projects normally entailed a considrable degree of uncertainty (p. 16), but failure to complete rojects on time was more often the result of diverting anpower to other endeavors or changing project objecives than a matter of unforeseen technical difficulties p. 200).

Despite uncertainties and slippage, Mansfield concluded hat a firm's expenditures for R&D were closely related to e total number of important inventions it produced and at innovations had demonstrable effects on the firm's rowth rate (pp. 199, 204). In recent years, spending for &D appears to be on the rise again, led by a number of ace-setting companies that are emphasizing technology nd innovation (Abernathy and Rosenbloom 1982, p. 418). otal expenditures for R&D by U.S. industries rose from 33.17 billion in 1978 to an estimated $49.15 billion in 1981, f which $1.55 billion was spent on basic research, $9.35

billion on applied research, and $38.25 billion on development (Peters and Fusfeld 1983, p. 13).

Recent studies of industrial innovation have stressed both the role of individual project champions and the managerial and organizational climate of the corporation. In ar examination of 73 case histories of innovations that 50 large companies had brought to successful commercialization, Fernelius and Waldo (1980) asked respondents to indicate the organizational and technical factors that affected their success. (The authors' rankings of these factors are shown in table 1.) Their interpretation of the data highlights the role of an individual in recognizing scientific technical, and market opportunities. These successful cases almost always had a "project champion": "someone who thoroughly believes in the project, works hard at it, inspires others to do the same, and defends the project even to the point of risking his own standing" (p. 39).

An unusual amount of encouragement from the top contributed to the success of one-third of these projects, and

TABLE 1
RANKINGS OF ORGANIZATIONAL AND TECHNICAL FACTORS AFFECTING THE INNOVATIVE PROCESS

1. Recognition of technical opportunity by an individual
2. Internal basic research
3. Recognition of market opportunity by an individual
4. Flexibility of project goals
5. Support of top management
6. Congruence of project with corporate goals
7. Complexity of project
8. Degree of project leader's autonomy
9. Recognition of market opportunity by a group; external basic research
10. Response to change in government policy
11. Recognition of technical opportunity by a group
12. Extent of external communication
13. Effect of government regulations
14. Defensive R&D
15. Extent of internal communication
16. Degree of project's urgency
17. Effect of project's urgency
18. Degree of uncertainty about changing government policy

Source: Fernelius and Waldo 1980, p. 37.

upport from top management was evident in almost all of he successful cases (Fernelius and Waldo 1980, p. 39). Vhether as a positive influence, as in these success stories, r as an inhibitor of change, managerial orientation and rganizational practices are frequently cited in the litera-ure as important factors for innovation.

Given the uncertainty involved in decisions about alloca-ions for R&D, the backgrounds and value orientations of ey managers may prove decisive in the ways projects are ssessed (Gold, Rosegger, and Boylan 1980, p. 20; Mans-eld 1968, p. 172). At the executive level, an orientation oward short-term gains and quantifiable results can dis-ourage strategies for long-term technological superiority nd preclude investment in the development of new prod-cts or processes (Abernathy and Rosenbloom 1982, pp. 16–18; Hayes and Abernathy 1980, p. 70).

The fostering of "an entrepreneurial culture" in which isk taking and innovation can thrive is an essential ingredi-nt for "high-technology management" in the large corpo-ation (Maidique and Hayes 1983, pp. 14–17). The willing-ess to experiment and to risk failure is also a key attribute f America's most innovative companies (Peters and Vaterman 1982, p. 48). An established pattern of commit-nent to innovation and technological advance may help a rm to attract and hold outstanding engineers and other echnical personnel, just as the absence of such commit-nent can have the opposite effect (Finniston 1979, p. 26).

"Internal entrepreneurs" have many of the same charac-eristics as "spin-off entrepreneurs," and company policies miting opportunities for young risk takers often prompt hem to leave the large corporation (Roberts 1968, pp. 251–3). In the case of one large electronics firm in Massachu-etts, former employees started 39 smaller companies. everal years later, the total annual sales of the 32 surviv-ng firms were twice that of the parent company (Roberts 968, p. 252).

These facets of technological change within industrial orporations point to the importance of several factors: (1) ne amount and quality of research and development; (2) ccess to developments in science and technology that resent new opportunities for innovation; (3) the caliber nd training of people in the laboratory and in corporate nanagement; and (4) an organizational climate that encourages continuous innovation. These factors have obvious implications for academic training, in business administration as well as in science and engineering. With regard to the transfer of technology, the major implication is the need for faculty and administrators to appreciate the kinds of difficulties faced by researchers and managers in the competitive environment of the corporation. Consulting, industrial associates programs, and other efforts to increase communication and mutual understanding represent positive steps in that direction.

Innovation through Entrepreneurship

It is hard to generalize about the relationship between size of a firm and innovation; in fact, innovative and noninnovative companies exist among both large and small concerns (National Academy of Sciences 1978, pp. 29–30). Sometimes a smaller enterprise invents but cannot complete the process of innovation; in other cases, invention occurs in a larger organization but is carried to completion by a smaller one founded by former employees of the original company.

Nevertheless, small firms led by enterprising entrepreneurs have an impressive record of producing innovations and creating new jobs. Local development programs to encourage new company formations represent "a relatively low-risk and potentially high-gain" approach (Shapero 1982a, p. 17). Rather than trying to attract the branch operations of established corporations, local and state efforts might be more successful if they concentrate on new firms and on the conditions conducive to their formation and development (p. 20).

What are the job-generating characteristics of small companies? Change in employment occurs through the birth, death, expansion, contraction, in-migration, and out-migration of companies (Birch 1979, pp. 3–6). Because about 8 percent of the annual job loss rate is the result of the death and contraction of companies, a key question is what kind of firms contribute to replacing these jobs. The start of new companies and expansion of existing ones are the major sources of replacement, and that replacement is related to the size and age of companies (Birch 1979).

- In Birch's study, small firms with 20 or fewer employees generated 66 percent of all new jobs, while middle-sized and large firms, on balance, provided relatively few new jobs.
- About 80 percent of all new jobs originate in firms that have been in existence for four years or less.
- The "job-generating firm" tends to be small, young, and dynamic.
- Small companies have higher death rates than larger concerns, but the surviving ones are four times more likely to expand than to contract, whereas larger firms are 50 percent more likely to shrink than to grow.
- These small, young, entrepreneurial firms are the kind "that banks feel very uncomfortable about" and are "the most difficult to reach through conventional policy initiatives" (pp. 8–17).

Small firms led by enterprising entrepreneurs have an impressive record of producing innovations and creating new jobs.

Such small, young, entrepreneurial firms are often associated with high-technology activities (Holtzman 1983, p. ; Joint Economic Committee 1982, pp. 19–21). But in a ample of spin-off firms in Michigan, a "continuum of technical information" was actually transferred when entrepreneurs left parent organizations to launch new companies, with university spin-offs concentrating chiefly on R&D, esting, and consulting and industry spin-offs tending to ocus on custom products and services (such as prototype lesign and fabrication) or on standard products and services (Lamont 1971, pp. 11–14, 30). High-technology companies actually constitute a very small fraction of the more than 14 million businesses in the United States (Grad and Shapero 1981, pp. 5–6), and high-technology activity represents less than one-third of the entrepreneurial growth of recent years, the major portion of which consists of services (restaurants, money market funds, and the like) and primary activities (education and training, health care, and nformation) (Drucker 1984, pp. 59–60).

Whether one prefers to think of the entrepreneur as anyone who starts any kind of new business or as any innovaor who champions a new technology, either in an existing orporation or in a new firm, many observers agree trongly that certain skills different from those involved in nvention are necessary to make a new business succeed.

Apart from technical considerations, a number of factors have been correlated with high performance in sales and profitability:

- **Moderate educational level:** The entrepreneurs possessing a Masters degree rather than a Ph.D. were more likely to be successful.
- **Specific business function:** The need for management skills was recognized and acted upon in forming the management team in successful companies.
- **Entrepreneur's concern about personnel matters:** The high performers tended to regard employees as the principal productive element of the company.
- **Marketing department:** Successful firms recognized the need to address market issues through a formal marketing structure (Roberts 1968, pp. 263–64).

Although reliable data of any kind on small firms are hard to come by, estimates of the number of small companies started vary from fewer than 500,000 to over 1 million per year. IRS records for 1965 to 1975 indicate that 2,563,000 firms net were added to the national inventory of businesses during that period (Grad and Shapero 1981, p. 14), and of all the new business firms started each year, only about 20 percent survive for five years (Johnson 1978 p. 11). Even so, small companies represent 95 percent of the businesses in the United States, employ 56 percent of the private, nonfarm work force, and account for 48 percent of the gross national product (Johnson 1978, p. 11).

High rates of failure in this important part of the economy have engendered widespread debate over ways to assist small businesses. Assessments of their needs usually cite managerial deficiencies, such as poor planning, inadequate controls, and insufficient understanding of finance. Such shortcomings may be the result of a lack of business training and experience (Brophy 1974, p. 182; Grad and Shapero 1981, p. 14) or of preoccupation with technological rather than business matters (Park 1983, p. 40). Entrepreneurs, especially technically oriented entrepreneurs, may not fully appreciate the importance of marketing skills and strategies (Lamont 1971, p. 41). The most severe problem of all may be obtaining capital to bring the business into being, establish a foothold, and support expansion (Shapero 1982a, p. 20).

The need for capital has been described as a "chicken and egg problem": "New firms will attract funds if they are successful, but they must have funds if they are to be successful" (Brophy 1974, p. 189). Although considerable attention has been given to the role of venture capital, much of this type of financing appears to be focused on technology-based enterprises concentrated in such states as California, Massachusetts, New York, and Texas (Pratt 1982, pp. 7–12). Moreover, venture capital is more readily available at a somewhat advanced stage of a company's development, whereas "seed capital" and "first-stage financing" to assess feasibility, develop a prototype, and then begin mass production and marketing are both more critical and harder to obtain (National Governors Association 1983, p. 16).

All of these obstacles must be overcome if small firms are to succeed and to make their maximum contribution to the economy. In terms of technological development, even the most well-conceived inventions may never be commercialized if entrepreneurs lack the necessary skills and resources to implement them (National Governors Association 1983, pp. 19–20). In terms of generating new jobs, entrepreneurial success in new and in expanded businesses may hold the key to overall employment growth in the nation (Birch 1979, p. 4). Assisting the small business sector holds a special challenge:

> *The firms that [economic development] efforts must reach are the most difficult to identify and the most difficult to work with. They are small. They tend to be independent. They are volatile. The very spirit that gives them their vitality and job generating powers is the same spirit that makes them unpromising partners for the development administrator* (Birch 1979, p. 20).

Insofar as they relate to small businesses, efforts to transfer technology in the strictest sense have usually focused on spin-offs and new, technology-based firms. But even in these cases, the needs of entrepreneurs go well beyond technical matters to such questions as the availability of capital, the provision of space and facilities, and all the other problems attendant upon operating a new business.

Cooperative Mechanisms for Technology Transfer

The classification of mechanisms for technology transfer is less standardized than for cooperative research models, which may reflect a shorter history of concern with transfer per se as well as the diverse factors involved in the process of technological change. Although definitions and terminology tend to vary, the major mechanisms for technology transfer include seminars, speakers, and publications, consulting relationships, industrial associates programs, extension services, industrial incubators and parks, and cooperative entrepreneurial development. What these mechanisms have in common are activities that provide information, technical services related to technological development, and/or managerial assistance.

Seminars, speakers, and publications

Faculty members play an important role not only as educators but as "skilled manipulators and suppliers of knowledge"(Bugliarello and Simon 1976, p. 3). Often these skills are tapped in informal, haphazard ways, as, for example, when professors receive telephone calls with requests for miscellaneous kinds of information or questions about various technical problems. More formal structures, such as technology clearinghouses or designated regional information centers, could make more effective use of faculty expertise (p. 79). Scholarly journals, interdisciplinary in nature and attuned to industrial needs, might be one means of communication associated with such programs (p. 85).

Of course, most academic departments in the sciences and in engineering do structure information services to some extent through conferences, speakers programs, and publications, especially for centers and institutes with a specific research focus and with an interest in advertising their programs. Corporations and industrial societies may also sponsor such activities. One chemical company, for example, arranges for faculty to serve as panelists in roundtable discussions, promotes seminars and speakers programs on new corporate developments, and underwrites distribution of a catalog of publications in its field of operation (Brodsky, Kaufman, and Tooker 1980, p. 71).

Periodic seminars and short courses are methods of technology transfer that promote "meaningful communication" and direct contact "where the action is: engineer to engineer, scientist to scientist" (Battenburg 1980, p. 8). Stepped-up efforts to disseminate information on entrepreneurship through journals, conferences, clearinghouses, and other measures would foster knowledge of entrepreneurs and their innovative role in the economy (Grad and Shapero 1981, pp. 34–36). About 40 percent of the institutions in one survey now publish some kind of journal or newsletter dealing with industrial development or regional economic development, and many of them plan to increase such publications in the future (Ferrari 1984, pp. 11–12).

Consulting relationships

Faculty consulting in industry has been described as "the most pervasive academic-industrial connection" (National Science Foundation 1982, p. 11). It is also a major element in the application of academic knowledge to mission-oriented problems in the industrial sector (Grad and Shapero 1981, p. 36).

> *Consulting is a touchy issue in universities, and is the subject of administrative regulation and criticism. It is . . . seldom discussed and relatively unmeasured, yet consulting probably does much to condition the viewpoints of professors vis-à-vis research and application of knowledge* (Grad and Shapero 1981, p. 37).

The extent of faculty consulting in industry is not precisely known. Many institutions do not have formal reporting requirements, and those that do seldom enforce them rigorously (Peters and Fusfeld 1983, p. 89). It is difficult to disaggregate the available data by type of income-producing activity and by type of client. Ninety percent of all faculty with nine-month appointments at four-year institutions earn some supplemental income, but that income includes summer teaching, research, royalties, and other sources in addition to consulting (Dillon 1982, p. 27). Survey data in various sources suggest that the proportion of faculty who engage in consulting may range from 40 to 49 percent (Brodsky, Kaufman, and Tooker 1980, p. 65; Dillon 1982, p. 38; National Science Foundation 1982, p. 11). External work as paid consultants differs markedly by type of institution and by academic field. University faculty do more than faculty at four-year or two-year colleges, and faculty in engineering and business do more than those in the physical and biological sciences (National Science Foundation 1982, pp. 11–14).

Institutional policies reflect a wide range of attitudes toward consulting: Some colleges and universities frown on it, others maintain a hands-off approach, and still others encourage it by such means as maintaining an inventory of areas of expertise and research interests. Reasons for promoting this type of link with industry may include supplementing faculty income, attracting research contracts, or maintaining a communications network (Peters and Fusfeld 1983, p. 89). Exposure to industry for students is another important objective, and some institutions have designed participation projects involving both students and faculty (Battenburg 1980, p. 8; Brodsky, Kaufman, and Tooker 1980, pp. 65–66).

From the viewpoint of corporate research directors, consulting may be perceived as a "means of facilitating general knowledge transfer from the university to their R&D staff," with the special advantage that academic researchers can be brought in on short notice, without the extended commitments required in contracted research (Brodsky, Kaufman, and Tooker 1980, p. 65). The single channel of commercialization used most often in one study was that of transfer to companies with which the university had a consulting relationship, especially if the association was a long-term one (Roberts and Peters 1981, pp. 123–24).

Faculty involvement in firms that extends to the point of participation in management or ownership of substantial financial holdings may be difficult to reconcile with academic status. The conflicts of interest that can arise in such cases have serious implications for the professor and for the institution. This problem is beginning to receive careful attention by academic leaders (see the following section). Even when faculty involvement is less extensive, a balance must be maintained between consulting activities and responsibilities for teaching and research.

Like other mechanisms for technology transfer, consulting can be viewed in various ways, depending on the aspect of technological development one is considering. In relation to established corporations, long-term relationships are more likely to be productive but are not always possible to achieve. Unless a cooperative research arrangement emerges, the link is individual rather than systemic (Brodsky, Kaufman, and Tooker 1980, p. 66). In relation to the encouragement of spin-off companies, the amount of consulting time allowed faculty may be far too limited (Southern Regional Education Board 1983b, p. 6). In relation to entrepreneurial firms in general, the present practice of consulting "predominantly for the government and very large firms in a few industries" might be augmented by "circuit-riding consultancies" targeted to small firms (Grad and Shapero 1981, p. 37). Such an approach, by which faculty members would be paid on the basis of the number of cases and small companies they handle, would be the small-firm equivalent of summer employment in corporations and would overcome the difficulty of waiting for problems to be brought to the faculty (pp. 37–38).

Instances of "institutional consulting" programs are rare (Peters and Fusfeld 1983, p. 92). To "institutionalize" consulting activities by faculty in engineering, business, and other areas, the more widespread adoption of professional practice plans similar to those employed by many medical schools might be necessary (Linnell 1982b, pp. 130–31). Such plans would have several advantages: selection of projects on the basis of professional interests rather than monetary factors, elimination of conflicts of interest, generation of income for the institution, and a more systemic rather than individualized link with industrial organizations.

Industrial associates programs

Another mechanism for transferring knowledge and fostering academic/industrial communication in areas of technology is the industrial associates program. Alternately called "liaison" or "affiliates" programs, these efforts help create a stable base of industrial support for academic research and provide participating companies with what academics like to call "a window on technology."

As with many of the transfer models identified in this report, the Massachusetts Institute of Technology and Stanford University have been the flag bearers, both having initiated associates programs many years ago. The MIT program has over 200 companies as members and is staffed by 11 liaison officers. Activities include symposia, seminars, visits to the campus, visits to the companies, and listings and reports on current MIT research (Bruce and Tamaribuchi 1981). Stanford has about 19 separate programs, because industrial membership and academic coordination occur at the departmental level. Faculty rather than staff members coordinate these efforts, each corporate member being assigned to a professor. The emphasis is on individual contacts, discussion of specific subject areas, and access to students—"the prime reason why companies join" (Peters and Fusfeld 1983, p. 80).

Many other institutions have associates programs—the University of California at Los Angeles, Cornell, the University of Washington, the University of Southern California, the California Institute of Technology, the Oregon Graduate Center, Lehigh University, and the Pennsylvania State University, to name a few (Battenburg 1980, p. 8; Brodsky, Kaufman, and Tooker 1980, p. 45). Some of these programs are institutionwide; others are departmental, frequently within schools of engineering.

Industrial representatives are exhibiting a growing tendency to question whether "general associates programs"—as distinguished from "special purpose" or "focused" programs—are sufficiently beneficial to justify the high membership fees, which often run to $25,000 or more. In fact, several institutions have launched general programs, only to see them fail for lack of industrial response (Peters and Fusfeld 1983, pp. 92–93). To provide corporations with regular technology overviews of sufficient depth across a range of areas is simply not possible unless an institution has a large and diversified research program.

Special-purpose associates programs, in contrast, seem to be growing in popularity. Membership fees are often considerably less, and the focus on specific subject areas may generate lively exchange between companies and university researchers. Industrial affiliates are encouraged to discuss nonproprietary technical problems with faculty members and suggest areas of investigation that might benefit industry. Such focused programs often evolve into research consortia (Peters and Fusfeld 1983, p. 79). The exchange of ideas on industrial problems may also lead to consulting relationships for participating faculty members (Smith and Karlesky 1977, p. 73).

As a model of technology transfer, associates programs tend to overlap with other mechanisms. They usually incorporate seminars, speakers, and publications within their array of services, and they may in turn be part of more inclusive programs like consortia or cooperative research centers. They may provide consulting services or lead to consulting relationships. And they have an educational or human resource development function as well, given the interest of corporations in access to students as potential future employees. The target of associates programs is typically the large corporation, though some institutions have experimented with sliding fee structures to accommodate smaller companies.

Extension services

Proposals dating back to the 1960s have attempted to promote innovation by linking university-based technology centers to federal programs to assist small companies (Baer 1980, p. 21). These "industrial extension services" are often advocated as offering to industrial firms the same kinds of benefits that agricultural extension has brought to the farmer. This rationale was behind the engineering experiment station pioneered at the University of Illinois in 1903 and expanded to 38 land grant colleges by 1937 (Peters and Fusfeld 1983, p. 98).

Several kinds of programs mentioned in the literature resemble, to various degrees, the agricultural model. In addition to the term "extension," one finds references to innovation centers, small business development centers, and other forms of "entrepreneurial assistance." "Extension services" seems to be a sufficiently inclusive umbrella term for such programs, all of which provide information and technical or managerial assistance, primarily to small businesses.

Some programs designed specifically to transfer technology to small firms borrow the terminology used in agriculture, and a county agent or extension agent plays a central role in transfer. Thus, the Ohio Technology Transfer Organization (OTTO) uses "technology transfer agents" situated at 11 community and technical colleges to provide technical assistance as well as information and training to small businesses throughout the state. The link to new technology occurs by networking the two-year colleges with a central office at Ohio State University, which maintains computerized data banks and has access to services of other state and federal programs (Warmbrod, Persavich, and L'Angelle 1981, p. 95). In Pennsylvania, "extension agents" are located at 24 continuing education offices as part of the Pennsylvania Technical Assistance Program (PENNTAP). Inquiries directed to PENNTAP are forwarded to "technical extension" experts in university departments, where faculty resources are tapped to provide the information needed (General Accounting Office 1983, p. 37).

Innovation centers and small business development centers (SBDCs), though often listed as separate models, have many characteristics common to all extension programs. The emphasis differs somewhat—innovation centers tend to focus on technological entrepreneurship, SBDCs on managerial problems of small businesses in general—but both types of programs offer entrepreneurs an array of direct services and frequently provide related education and training as well.

In 1973, the National Science Foundation funded three innovation centers—at M.I.T., Carnegie-Mellon, and the University of Oregon. The objective was to combine classroom training in engineering and business theory with hands-on, clinical experience in generating new ideas, developing new products, and initiating new ventures. By 1978, over 1,000 students had enrolled in 25 "new venture courses," 26 new businesses had been launched with projected gross sales of over $12.5 million, and nearly 800 new jobs had been created (Colton 1978, pp. 193–94). These centers, and numerous others developed at other institutions, integrate formal evaluations of ideas submitted by independent inventors and existing businesses into the training program, then provide support services as the new ventures begin to take shape. The merit of such programs lies in their attempt to deal with the full range of problems encountered from invention through implementation:

> *By spanning the entire innovation process from the generation of new ideas to the actual manufacturing and marketing of a new product, this mechanism seems to offer an ongoing, flexible support system for young entrepreneurs* (Brodsky, Kaufman, and Tooker 1980, pp. 48–49).

Small business development centers have been established at many academic institutions. SBDCs within universities are usually situated in schools of business administration, and a number of states also have statewide programs. In Georgia, 150 people staff 11 centers in various sites. Partially funded by the U. S. Small Business Administration, these centers provide consultation, continuing education, and special programs on international trade, economic forecasting, energy management, and minority business concerns (Georgia Office of Planning and Budget 1982, p. 33).

Such programs often attempt to collate information on a variety of business matters of special interest to owners of small firms, such as the availability of investment capital or of governmental financing. At least six federal agencies have small business assistance programs, the newest of which is the Small Business Innovation Research Program, legislation for which was enacted in 1982. Legislation creating the program requires certain agencies to set aside 1 percent of their research and development budgets for small businesses. Several states—Nebraska, North Carolina, and Pennsylvania, for example—have mounted efforts to bring representatives of academic institutions, small business, and financial institutions together to take advantage of this program (National Governors Association 1983, pp. 25–26).

Although few of these extension programs directed toward industry are "full blown replicas of the agricultural extension model" (General Accounting Office 1983, p. 36), such programs "do establish a network of industrial contacts and make the universities that participate more sensitive to industrial needs" (National Science Foundation 1982, p. 23).

Industrial incubators and parks

Over the years, a number of academic institutions interested in closer links with industrial firms have set up or participated in arrangements to provide physical facilities for companies. Often an outgrowth of prior technology transfer or technical assistance, such arrangements further these efforts through geographical proximity. The two major forms are industrial incubators and industrial parks.

The objective of industrial incubators is "the creation of an interactive environment between industry and education" (*Venture Capital Journal* 1983, p. 9). One of the oldest incubators is that of the University City Science Center in Philadelphia, a joint venture of several colleges, universities, and local industries founded nearly two decades ago. Other well-known incubator programs are located at the Georgia Institute of Technology and at Rensselaer Polytechnic Institute (RPI).

Georgia Tech's Advanced Technology Development Center provides both technical and management advice to small firms that occupy space in a 100,000–square foot facility. Access to faculty consultants, equipment, and library and computer services helps entrepreneurs develop to the stage where they are able to set up their own plants (Georgia Office of Planning and Budget 1982, pp. 18–19). At RPI, similar services are offered in low-rent space on the campus as part of an institutionwide effort to foster high-technology entrepreneurship. The main criteria for the selection of tenants are the marketability of their proposals and correspondence to RPI expertise. Occupants may be inventors, faculty members, or students. Several fledgling companies are already producing pharmaceuticals, solar collectors, robot control systems, and automated test equipment. Although it is hoped that some of these organizations will move to RPI's new, off-campus industrial park, none had yet done so by the end of 1983 (Phalon 1983, p. 91).

Whether or not the industrial or research park actually represents "the most dramatic contribution to innovation" of the various forms of technology transfer (General Accounting Office 1983, p. 48), it is certainly the most visible. Precisely because of its concreteness, it is often the image that Chamber of Commerce members and university trustees have in mind when thinking about academic/industrial cooperation. A park site close to an academic institution breaks down spatial barriers and may thereby make it easier for academic and industrial researchers to interact more frequently and intensely, share each others' facilities, and develop cooperative programs.

Originally called "industrial estates" in the United Kingdom and then "industrial districts" in this country, such sites are now variously termed industrial parks, research parks, or science parks. By 1961, 1,046 known "industrial districts" existed in the United States and Canada (Lee 1982, p. 34). The first site established by a university was the Stanford Industrial Park, founded in 1951 on land adjacent to Stanford University. One of the newest is the Science Park in New Haven, jointly sponsored by Yale University, the Olin Corporation, and the city of New Haven.

The park site model is treated these days with caution—if not disapproval—by most careful observers.

Despite the success of Stanford's park in spawning new companies and of North Carolina's Research Triangle Park in attracting governmental and corporate R&D facilities, the park site model is treated these days with caution—if not disapproval—by most careful observers. Several studies in the 1960s and 1970s reported that more research parks were failing than succeeding (Carter 1978, p. 1470; Lee 1982, p. 3). Although the exemplary cases are generally considered to be excellent modes of transferring technology, recent commentaries tend to emphasize the difficulty of making them work (General Accounting Office 1983, p. 50; Joint Economic Committee 1982, p. 42; National Governors Association 1983, p. 17).

Parks developed by universities or by private developers acting unilaterally have not fared well (Southern Regional Education Board 1983b, p. 6). Certain conditions, missing in one-sided arrangements, are necessary for success:

> *The probabilities of success for these ventures increase dramatically when a communitywide, diversified approach is taken, involving active participation by the university, private developers, representatives of local high-technology industries, and community leaders.*

When any of these is missing, the chances for success decline rapidly (Southern Regional Education Board 1983b, p. 6).

A number of strategic decisions, if made incorrectly, can diminish the impact of parks on technological development (Lee 1982, pp. 116–29). Strong commitment by the university to the industrial development needs of the region is critical, however.

Although industrial parks afford numerous benefits to the university, such as generating income and increasing employment opportunities for graduates, the self-interested objectives of the university tend to fold into the broader ones of economic development in the community. The academic institution is thus challenged to become an active partner with both public and private sector organizations for economic improvement.

Cooperative entrepreneurial development

The sixth model of technology transfer is but partially and imperfectly reflected in the literature. As noted earlier, these models often overlap, and as the points of contact increase and objectives multiply in the interest of technology transfer, the range and intensity of academic involvement with industrial concerns and sometimes with groups in the community increase as well. The research park illustrates this concept well, with success or failure depending on cooperation from many organizations, both public and private.

The term "cooperative entrepreneurial development" is used here to reflect a more comprehensive approach to promoting the transfer of technology and the development of entrepreneurship through a broad range of cooperative activities focused on the community or region. While few if any perfect examples of it can be found, its components can be identified and illustrated.

One loose category of activities related to entrepreneurial development is what is sometimes called "technology brokering." A technology broker is a person or agency whose role it is "to bring universities with research capabilities together with industrial firms with research needs" (Baer 1977, pp. 49–50). Four types of brokers have been identified: (1) university foundations; (2) independent groups, such as the Battelle Development Corporation or the Research Corporation; (3) private consulting firms; and (4) government agencies (Baer 1977). From a university's viewpoint, such third-party mechanisms create a neutral buffer between academic activities and business dealings, provide professional brokerage expertise, generate continuing income for the institution, and establish a structure with which industry can relate (Peters and Fusfeld 1983, p. 111).

As part of their brokering, some institutions have internal programs to identify and encourage new technologies with commercial potential. For example, the program at Case Western Reserve University (called "Quest for Technology"), with the assistance of the Control Data Corporation, invites faculty and students to submit ideas for machines, computer programs, new applications of existing substances, training devices, and the like. The inventions that are accepted are then marketed to investors willing to provide financing for business start-ups (Wood 1983, p. 60).

Some universities have taken a further step: to participate in the financing of entrepreneurs. Although it is increasingly common for institutions with substantial endowments to include venture capital companies in their investment portfolios, financing spin-offs from the university itself is a new idea (*Venture Capital Journal* 1983, p. 11). Thus far, little information is available on how many institutions are now involved in or may be considering equity participation, the "closest and most controversial of all university/industry ties" (National Governors Association 1983, p. 14). Nor is it clear that the advantages to be gained by financial commitments to entrepreneurs within the university outweigh the very real dangers to academic integrity (see following section).

In addition to technology brokering, cooperative entrepreneurial development is also partially illustrated in the advanced technology application centers in states like Pennsylvania and Ohio. In both of these states, areas designated as technology application centers are expected to engage in cooperative activities referred to variously as "technology transfer" or "entrepreneurial assistance" programs. The concept is to link university-based R&D pursuits with cooperative efforts in technical and business assistance, incubator services, investment opportunities, and other community resources (Ohio Department of Development 1983, pp. 4–5; Pennsylvania Department of Commerce 1983, p. 1). By attempting to bring the major academic, industrial, and governmental organizations in an area together as sponsors of and participants in the whole array of activities encompassing technology transfer, the advanced technology application centers probably represent the best examples of cooperative entrepreneurial development that can be found at this time. As experience with such programs increases, their multiple objectives will likely attain sharper focus, and clearer examples of this model will emerge.

Beyond Technology Transfer

In these mechanisms of technology transfer, it is evident that much more is occurring in many instances than merely the transmittal of information. Just as the models of research cooperation reflect a spectrum of complexity, so do these transfer models. This conclusion seems inevitable, given the complexity of the innovation process itself and the multiple factors and target audiences involved. An interest in "integrating technological results of university research into private sector programs or commercial products" (Peters and Fusfeld 1983, p. 98) quickly leads beyond the traditional mechanisms of informational events and publications, consulting, and associates programs to more ambitious efforts to provide extension programs, incubator facilities, park sites, and accompanying services to corporations and small businesses.

If the literature on technology transfer exhibits a certain indefiniteness, it is doubtless a result of the difficulty in simultaneously keeping in focus the innovative requirements of existing corporations and of small entrepreneurial firms. The small companies, moreover, differ in the degree to which technology is central to their operations. And in all cases, it has become evident from a review of the variables that influence innovation that many factors, including managerial needs, are as important as purely technical matters.

In short, the term "technology transfer" hardly seems elastic enough to carry all the weight associated with these multiple activities and relationships. The term "entrepreneurial development," though not without its own difficulties, may be more descriptive. While some writers prefer to associate entrepreneurship with any kind of small, new business (Drucker 1984; Grad and Shapero 1981), a long tradition of identifying entrepreneurial behavior within the corporation as the bearer of innovation is also apparent. "Clearly, the entrepreneur is the central figure in successful technological innovation, both within the large corporation as well as the foundling enterprise" (Roberts 1968, p. 259). If this view of entrepreneurship as applicable to both large and small concerns is followed, it is still important to remember that non-technology-based firms are also embraced by the term.

Summary

This discussion of technology transfer has been premised on the fact that, from the viewpoint of economic development, research and development is not an end in itself but a means to stimulate technological innovation. When the process of innovation is viewed in its entirety, however, it becomes apparent that the concern to exploit new concepts that may derive from academic research necessitates attention to the informational and decision-making needs of the corporation, the technical and managerial problems of small firms, and the environmental factors that exist in the community setting.

The transfer mechanisms described in this chapter represent attempts to address these needs. If no single model does so completely, all of them taken together reveal an interest on the part of academic institutions, industrial organizations, and other agencies in many parts of the country to create the conditions necessary to complete the cycle of innovation leading from invention to successful commercialization. The role of innovators and entrepreneurs, whether in large companies or small, is widely regarded as critical to accomplishing this goal, and an increasingly intricate network of services and relationships has evolved to assist in entrepreneurial development. In many cases, these efforts go far beyond technology transfer in the narrowest sense of transmitting information.

It is clear that the propulsion of the processes of technological change (invention, innovation, and diffusion) is a highly complex matter, not yet fully understood by those who study it and attempt to engage in it (Brophy 1974, p. 180).

Yet the failure of a community or region to identify technology-based market opportunities, promote relevant R&D, and stimulate the transfer of technology from invention through diffusion does have predictable results in the form of technology-lag and economic decline (Brophy 1974).

ACADEMIC/INDUSTRIAL COOPERATION IN PERSPECTIVE: Major Dimensions, Concepts, and Issues

Properly understood, technology (and innovation) encompasses the organizational setting in which tools are deployed, the work roles of people involved in their use, and the perceptions of actors involved in adoption and implementation (Tornatzky et al. 1983, p. 14).

The high-technology connection in which higher education and industry are joined to stimulate technological progress has been discussed thus far in terms mirroring the dominant themes in the literature. These themes emphasize the exciting potential of advanced technology for economic growth, the role of America's leading research universities in advancing scientific and technological frontiers, and the primary forms of university cooperation with industry in R&D and in technology transfer.

Exploration of these themes, however, has revealed additional forces shaping academic/industrial relationships. To put these relationships into broader perspective, it is necessary to consider human resources, another important part of comprehensive development strategies, to consider a conceptual framework that encompasses all of the major areas of cooperation and relates them to the main goals of economic development, to review the barriers to cooperation and ways of analyzing cooperative interactions, and to describe some of the major policy issues that arise as academic and industrial organizations move toward closer alliances.

Recognizing the Dimension of Human Resources

As readers interested in other dimensions of academic/industrial relations will have noticed, the paradigm for the material reviewed up to now derives from the research milieu of the university. As new technology depends on fundamental investigation, the research paradigm is a useful and appropriate perspective from which to examine the process of technological innovation. It is not the only perspective available, however, nor is it without its disadvantages.

Some of the limitations that may result from reliance on the research paradigm have already been alluded to in previous chapters:

1. An overemphasis on the roles of universities, especially the top 100 or so, as opposed to the full range of academic institutions in this country, which total over 3,000;
2. An overemphasis on large corporations, especially those Fortune 500 companies with substantial R&D, rather than the wide diversity of business enterprises numbering in the millions;
3. A tendency in some cases to view cooperative research projects as support mechanisms rather than partnerships in technological development;
4. The difficulty of addressing the myriad needs for support of entrepreneurship within the conceptual parameters of technology transfer;
5. A tendency to neglect the development of human resources, an extremely critical component of technological development; or, when education is noted, a tendency to overemphasize the level of advanced study associated with research rather than the full spectrum of educational and training needs.

This paradigm, of course, is not the ivory tower perspective that eschews all interest in the use of research findings and rejects close relationships with other organizations in favor of isolated scholarship. Although the metaphor of the ivory tower persists (Crosson 1983, p. 10), the question for most educators today is not whether higher education should be involved in society but how to meet social responsibilities and still fulfill the academy's mission (Ashby 1958; Bok 1982; Kerr 1972; Millett 1968). Yet the research paradigm in the literature on academic/industrial relationships does take academic research as its starting point and differs, for example, from the viewpoint of those who look at the matter from the perspective of human resources.

Since the days of Adam Smith, human resource development, "the process of increasing the knowledge, the skills and the capacities of all the people in a society" (Harbison and Myers 1964, p. 2), has been recognized as critical for the development of national economies. A sizable body of literature deals with the dimension of human resources in academic/industrial relations, though seldom comprehensively. Three aspects of that literature are especially important for economic development: scientific and engineer ing manpower, the need for training and retraining, and information about the labor market and occupational forecasting.

Scientific and engineering manpower
The total number of faculty in science and engineering in public and private universities offering the doctorate is now about 220,000, but the full-time equivalent number engaged in research and development in these institutions is only about 58,000 (Baker 1983, p. 113). Fewer than half this number are in the physical sciences and engineering. Depicting this situation as a massive problem of "understaffing," Baker declares that it is an "illusion" to think that academic/industrial cooperation can significantly improve America's innovative capacities without a major increase in scale.

Actually, the academic employment of scientists and engineers has been increasing at about 3 percent annually in recent years (National Science Foundation 1980, p. 1). Much of this growth, however, has occurred through hiring non-tenure-track research staff on short-term contracts. Because of projected declines in enrollment, these "soft-money" positions may be vulnerable at institutions other than the most distinguished universities (National Science Foundation 1980). Lower enrollments may reduce academic opportunities for young scientists and lead to shortages of research personnel in the years ahead.

In engineering colleges at present, course offerings are being reduced and research effort decreased as a result of insufficient faculty. About 1,650 positions are vacant across the country, and heavy teaching loads and obsolete instrumentation are commonplace (National Governors Association 1983, p. 10). Faculty shortages are especially acute in specialized fields like computer science, computer engineering, robotics, and CAD/CAM (Geils 1983, p. 49).

Among high-technology companies in the Silicon Valley and along Route 128, an inadequate supply of graduates in science and engineering was a serious concern of company officials (Useem 1981, pp. 19–20; 1982, p. ii). Consequently, access to capable graduate students frequently turns out to be a major motivator for corporate participation in cooperative programs with universities (Cromie 1983, pp. 245–49). In fact, the supply of high-level scientists and engineers to industry may be the single most important incentive for academic/industrial research links (National Science Foundation 1982, pp. 29–30; Peters and Fusfeld 1983, p. 93; Shapero 1979, p. 4).

Training and retraining

Yesterday's futuristic prophecies—the electronic office, the automated factory, the computerized household—are rapidly becoming today's realities. An estimated 55 percent of the workforce is already employed in information/knowledge industries (Jamieson and Warren 1980, pp. 18–20). This kind of work demands verbal and quantitative competencies, perhaps requiring total retraining and recertification of many workers. Over the next two decades, the skills of over 40 percent of the current workforce may become obsolete, pointing to the need for "training and retraining . . . to smooth the transition to a technology-based society" (National Governors Association 1983, p. 20).

Given the large number of unemployed workers in older industrial areas and the importance of skilled labor to high-technology companies, training is a key ingredient in state initiatives to promote technological development (Joint Economic Committee 1982, p. 39). Industry itself is spending substantial sums for training—from $10 billion to $30 billion or more per year (Honan 1982, p. 7). In some cases, difficulties in finding qualified applicants for technical positions have prompted corporate employers to initiate cooperative training programs with academic institutions, particularly community and technical colleges (Georgia Office of Planning and Budget 1982, pp. 45–46).

The nearly 1,300 two-year community and technical colleges in the United States represent an important resource in meeting the nation's training needs. When new technologies are introduced in industry—altering work settings, tools and processes, and performance requirements—training programs must be revised accordingly. To keep up with such changes, two-year colleges need to update curricula, acquire or gain access to state-of-the-art equipment, and locate qualified instructors (Long and Warmbrod 1982, pp. 1–2). To avoid recurrent obsolescence of programs, it is necessary to teach generic as well as specialized skills and to augment the credit curriculum with more flexible, tailor-made, noncredit offerings (Edling 1982, pp. 4–8).

Much of the current literature on training emphasizes that effective training strategies depend on joint corporate/collegiate planning and on communitywide coordination (American Association 1983, pp. 7, 17; Long and Warmbrod 1982, pp. 5–6; National Governors Association 1983, p. 21). Interest is also growing in statewide coordination. Although virtually all 50 states have manpower development programs, many of them featuring customized job training for new companies entering the area (Urban Institute 1983, p. 14), few if any states have comprehensive programs linking job training to the capabilities of their postsecondary education system (Wilson 1981, p. 11).

Over the next two decades, the skills of over 40 percent of the current workforce may become obsolete.

The labor market and occupational forecasting

A mismatch between job openings and the skills of the workforce is increasingly apparent in cities across the country. When high unemployment exists side by side with severe shortages of skilled personnel, it is evident that human resources are not being used to the fullest. An important economic objective, therefore, is to improve information about the labor market and manpower projections to guide education and training.

Industry's requirements for manpower have shifted toward higher levels of education, and the nation's educational institutions have responded with new programs and delivery systems—but in a "piecemeal and uncoordinated fashion" (Kyle 1981, p. 101). A number of reasons are apparent for the persistent imbalance, but the inadequacy of labor market information systems is one factor that can be corrected (Kyle 1981). The establishment of "human resources management centers" is one way to correct the imbalance: Colleges and universities, local businesses, and government agencies would join forces "to provide the information base and technical assistance needed to avoid serious disruptions in the local economy" (Kyle 1981, p. 102).

Information garnered from employers, however, does not always give clear signals on long-range requirements. Among technology-based companies in the Boston area, for example, some employers anticipated reduced needs for technicians as a result of automation, while others cited different factors likely to increase demands (Useem 1982, p. ii). These responses point to the more fundamental problem of interpreting complex and often countervailing trends in the economy: In which areas will employment gains be greatest? What importance should be attached to advanced technology, to the growth of service industries, or to the shift to an information society? A number of writers, extrapolating from past labor statistics, predict growth in high-technology employment but see it as a relatively small part of the overall future economy (Peterson 1982; Pollack 1984; Rumberger 1983). Other sources, taking more of a futurist approach, foresee vast changes in which new technology and people with the requisite skills will play a crucial role (Helms 1981, pp. 7–14; Molitor 1981, p. 23; Naisbitt 1982, pp. 49–52).

Does anyone really know what jobs will exist, say, in the year 2000? We do not presently have a system to forecast the occupations that will be created by technological change (Helms 1981, p. 15). Such a system would require tracking new developments in science and engineering, collecting and analyzing R&D data, studying new production facilities, processes, and products, and identifying changing requirements for skills and knowledge (pp. 17–18). Information collection and analysis of this kind are elements in the "emerging science of occupational forecasting" (p. 17), which Helms regards as essential to prepare people for productive roles in tomorrow's workforce.

These aspects of human resource development pertaining to professional manpower, training/retraining, and forecasting of new occupations and needed manpower all constitute additional agenda items for the high-technology connection. The major mechanisms for academic/industrial cooperation for the development of human resources include professional and technical degree programs, business/industry advisory committees, cooperative education, continuing education courses, extended degree programs, nontraditional credit programs, industrial adjunct faculty, and cooperative planning and program councils.

Scientific, engineering, business, and other professional and technical degree programs. Such institutional programs are found at various levels in most academic institutions; they represent those areas of the curriculum that are especially oriented toward business and industry. Faculty from these areas are generally most active in pursuing academic/industrial links (Ferrari 1984, p. 17). (For industrial evaluations of the graduates of these programs, see Lynton 1981, Sirbu et al. 1976, and Useem 1981, 1982.)

Business/industry advisory committees. Most institutions have departmental or collegiate advisory structures to gain input from the private sector. They are especially common in two-year colleges. The extent of actual collaboration between the institution and corporate representatives varies, however. Whether this mechanism provides a strong enough voice for industry is questionable (Cross 1981, p. 6). That its effectiveness depends on a genuine desire for advice rather than simply an interest in seeking financial support is apparent (Battenburg 1980, p. 9). Some corporations, particularly those heavily involved in R&D, have technical advisory boards of their own on which academic scientists sometimes serve (Baer 1977, p. 49).

Cooperative education. Extolled by many observers as one of the most effective of all models of academic/industrial cooperation, "co-op" provides students with opportunities to test career directions, institutions with a mechanism to integrate theory and practice, and employers with a low-cost method of identifying potential future employees (Wilson n.d.). Alternating periods of study and of work—called "sandwich courses" in the United Kingdom and simply "alternation" in Europe—has the potential of bringing institutions and industries closer (Organization for Economic Cooperation 1982, pp. 61–63). First established in this country at the University of Cincinnati in 1906, cooperative education grew slowly but steadily until the 1960s, when federal funds became available and encouraged expansion to over 1,000 colleges and universities (Brodsky, Kaufman, and Tooker 1980, p. 60).

Continuing education courses. Offerings in continuing education have grown steadily in recent years; they are increasingly important for updating the workforce at every level. The number of four-year colleges and universities that operate extension, continuing education, correspondence, and various noncredit programs was estimated at 1,233 in 1978, more than twice the number offering such programs in 1967–68 (Peterson 1979, pp. 24–25). Advocates of lifelong learning continue to press for the integration of continuing education into the mainstream of academic life (Votruba 1978). For high-technology industries, participation in continuing education is particularly important to avoid obsolescence in scientific and engineering knowledge (Brodsky, Kaufman, and Tooker 1980, p. 54). Some private colleges offer special training on managing high-technology enterprises to industry executives (Wood 1983, p. 62).

Extended degree programs. External or extended degree programs are awarded "on the basis of some program of preparation . . . not centered on traditional patterns of residential collegiate or university study" (Houle 1974, p. 15). Adults employed in industry or working at home usually constitute the student body, and access is provided by holding courses at different times or places or by using alternative modes of instruction (Johnson 1984, p. 484). A national survey disclosed 244 undergraduate external degree programs offered by 134 institutions in 1976 (Sosdian and Sharp 1977, p. vii). Such programs may be offered by single institutions or by a consortium of colleges (Valley 1979, pp. 156–73). In some cases, they are offered in conjunction with local industries to meet employees' needs through flexible scheduling, in-house classes, or use of televised instruction (Kyle 1981, pp. 105–7; Useem 1981, p. 21; Valley 1979, p. 176).

Nontraditional credit programs. Programs such as the American Council on Education's Program on Noncollegiate Sponsored Instruction (PONSI) and the Council for the Advancement of Experiential Learning (CAEL) have enlisted growing numbers of institutions in efforts to award credits for extrainstitutional learning that equates to academic coursework (Cross 1978, p. 23). The ACE program uses panels of experts to evaluate courses given in industrial and other nonacademic settings and to recommend amounts and types of academic credit that might be awarded to those enrolled. The experiential learning program coordinated by CAEL helps member institutions develop tools for assessing competencies gained through experience in the workplace or elsewhere. A study of 99 public and private colleges in Ohio found that a majority of responding institutions endorsed the awarding of credit for both noncollegiate and experiential learning under certain conditions (Cameron 1980, p. 2). Such efforts, though still controversial, can be viewed as important ways to bridge the gap between education and businesses and industries (DeMeester 1981, pp. 74–76).

Industrial adjunct faculty. The use of knowledgeable people from industry as adjunct faculty is a well-established mechanism of cooperation—one that is especially valuable for instruction in new areas of technology. Community colleges will find it increasingly necessary to use technical experts as part-time instructors, and full-time faculty may find it beneficial to attend their sessions as well (Edling 1982, pp. 4–5). The number of adjunct faculty appears to be increasing in engineering schools because of the shortage of regular faculty (Peters and Fusfeld 1983). The larger significance of adjunct professorships for academic/industrial cooperation undoubtedly depends on how the campus and contributing organization handle the arrangement.

Cooperative planning and program councils. Increasingly, colleges and universities are working with the private sector in cooperative planning and sponsorship of programs to meet changing needs for education and training (Craig and Evers 1981, pp. 41–42; Cross 1981, pp. 5–6; Warmbrod, Persavich, and L'Angelle 1981, pp. 63–67, 107–11). Community and technical colleges have often led the way in such partnerships, but private colleges and state universities are involved too, and sometimes a number of institutions of different types work together in a particular region or locality. Ongoing, systematic cooperation is often recommended to assess needs, inventory resources, and formulate respective roles, but such cooperation appears to be more the exception than the rule. Many opportunities, such as tuition aid programs, are still underused, and the needs of the workforce are larger than either sector can meet acting alone (Craig and Evers 1981, pp. 41, 43; Lynton 1981, pp. 10, 13–15).

This brief overview of education and training programs hardly does justice to the human resources dimension of the subject. It does suggest, however, the vast amount of joint activity in this area, which for many writers is the *main* arena of academic/industrial linkages. The extent of cooperative effort as well as its importance for economic development more than justifies the inclusion of human resource development in any conceptual framework of intersector relations.

Conceptualizing Academic/Industrial Cooperation

The literature on economic development may be the best place to look for guidance on the conceptualization of academic/industrial cooperation. A reading of recent development plans in practically any state will certainly disclose references to research and its link to advanced technology, but they appear alongside many other concerns. In Pennsylvania, for example, the Ben Franklin Partnership Program has three cornerstones:

- **Joint research and development,** in concert with the private sector, in specified areas like robotics, biotechnology, and CAD/CAM. (These areas vary by center, with each emphasizing four or five areas.)
- **Education and training,** assisting all institutions of higher education to train and retrain individuals in the skills essential in starting and expanding firms.
- **Entrepreneurial assistance services,** which include linking R&D, entrepreneurs, venture capitalists, and other financial resources; assisting in the preparation of business plans and feasibility studies; and providing small business incubator space and services and technology transfer (Pennsylvania Department of Commerce 1983, p. 1).

Here again are the three essential bases identified earlier that link academia to industrial and governmental efforts to promote technological development. Table 2 illustrates the substantive areas of interaction, summarizing relationships observed in the development literature between major economic goals involving technology and specific developmental strategies. All of these relationships assume cooperation between the higher education and corporate com-

TABLE 2
ACADEMIC/INDUSTRIAL RELATIONSHIPS IN RELATION TO ECONOMIC DEVELOPMENT GOALS AND STRATEGIES

Economic Goals	Developmental Strategies	Academic/Industrial Relationships
Generation and application of scientific/technological knowledge.	Strengthen basic and applied research in colleges and universities.	**RESEARCH AND DEVELOPMENT RELATIONSHIPS**
	Increase interaction among basic research, applied research, and development processes as they occur in academia and industry.	
Trained manpower at all levels for technological employment.	Strengthen scientific, engineering, business, and other professional and technical programs in higher education.	**HUMAN RESOURCE DEVELOPMENT RELATIONSHIPS**
	Strengthen training and retraining in technology skills.	
	Improve labor-market information and occupational forecasting.	
Effectiveness and innovativeness in new and existing industries.	Support technology transfer mechanisms to increase innovation among larger, established firms and small entrepreneurs.	**ENTREPRENEURIAL DEVELOPMENT RELATIONSHIPS**
	Encourage a variety of services to help entrepreneurs in creating and expanding new industries.	

munities and center around research and development, human resources, and entrepreneurial development.

In this broader conceptual framework, research still occupies an important position, but the relationships denoted as technology transfer, in keeping with the research paradigm, now are subsumed under the broader category of entrepreneurial development. Further, the interactions related to education and training now constitute a category of their own as human resource development relationships. The three major categories, it might be added, reflect the primary goals and strategies in current economic planning for technological development without distorting the major mission of higher education in research, teaching, and service.

Still another advantage of this framework is that it recognizes the potential roles of a larger number of participating institutions and firms. Whereas the research paradigm tends to highlight only the leading research universities and the largest industrial R&D performers, the model suggested here can be adapted to the whole panorama of collegiate and corporate organizations. Clearly many hundreds of colleges and companies can elect to work together to further one or more of the major goals of economic improvement if they desire to do so.

In addition to identifying the substantive areas of interaction, a conceptual framework should also speak to the qualitative nature of academic/industrial relations. Genuine cooperation in research or in sharing facilities is still "uncommon," most high-technology linkages are initiated by academic staff, and extending additional forms of service as in industrial incubator programs may generate more industry-initiated demands than academic institutions are prepared for (Southern Regional Education Board 1983b, p. 4). Further, intersector relationships have thus far had only a modest influence on the activities of academic and industrial organizations (Gold 1981). The vested interests of neither side have yet been engaged: "We have not yet reached a point where the enrollments of higher education or the profits of corporations have been tied to direct collaborative planning and action" (p. 13).

Table 3 incorporates the qualitative aspect of relationships by categorizing the major models according to three levels of interaction. At the lowest level of interaction are

TABLE 3
ACADEMIC/INDUSTRIAL RELATIONSHIPS BY TYPE AND LEVEL OF INTERACTION

LEVEL OF INTERACTION	MAJOR TYPES OF INTERACTION: **Research and Development Relationships**	**Human Resource Development Relationships**	**Entrepreneurial Development Relationships**
Academic/Industrial Partnership	Cooperative Research Centers	Cooperative Planning and Program Councils	Cooperative Entrepreneurial Development
Academic Activity in Collaboration with Industry	Research Consortia Personnel Exchange Programs Special Research Agreements Contract Research	Industrial Adjunct Faculty Nontraditional Credit Programs Extended Degree Programs Continuing Education Courses Cooperative Education Business/Industry Advisory Committees	Industrial Incubators and Parks Extension Services Industrial Associates Programs Consulting Relationships
Academic Activity Oriented toward Industry	Research Centers and Institutes	Scientific, Engineering, Business, and Other Professional and Technical Degree Programs	Seminars, Speakers, and Publications
MAJOR ECONOMIC GOALS	**Generation/Application of Knowledge**	**Trained Manpower**	**Industrial Innovativeness**

those academic activities, such as campus-based research centers, various degree programs, and publications or speakers programs, that are *oriented* toward industry but remain wholly within an institutional framework. The next level of interaction consists of *collaborative* activity with industry. Most linkage programs occur at this level: A significant degree of exchange occurs between academia and industry, without fundamental organizational change. The highest level of interaction is the *partnership*. Although the term "partnership" is used rather loosely in the literature, actual examples of this level of interaction are harder to find. "True partnerships" are marked by long-term, formal agreements, significant effort of mutual benefit, and joint planning, management, and implementation (Prager and Omenn 1980, p. 379). In a partnership, "the whole is greater than the sum of the parts."*

Few "pure" examples are available of any of these models, which often display different characteristics in different settings. Moreover, actual relationships tend to resist characterization along a single spectrum of levels of interaction. Nonetheless, real qualitative differences exist: Some models go much farther in genuine interorganizational sharing than others, and those at the upper end of the spectrum tend to display broader agendas of cooperation. Depending on one's viewpoint, the partnership is not necessarily the "best" or most desirable model. It may be the most effective (and also the most problematic) way, however, to achieve the desired outcomes of economic development.

Analyzing Interactions and Barriers

A British observer stated several years ago that " 'if the idea of collaboration between universities and industry is buried underneath sufficient platitudes, it will die of suffocation' " (D. C. Freshwater, cited in Baer 1977, p. 55). There is, indeed, a great deal more rhetoric in the literature on academic/industrial relationships than there are data and analyses of outcomes, and barriers to analyzing these data do exist. Generally, they are of three kinds: (1) limited resources; (2) organizational differences; and (3) organizational rigidities.

*Theodore Settle 1983, personal communication.

Limited resources
"Limits on available faculty time and limits on available industrial resources" (National Science Foundation 1982, p. 30) affect joint activity. Complex interorganizational arrangements are very demanding, especially those requiring the greatest amount of human input and facilities (Brodsky, Kaufman, and Tooker 1980, p. 79). The essential ingredient of scientific excellence is itself a limiting factor on the academic side, which accounts for the primary concern of many academic leaders for basic institutional support from state, federal, and other sources (Ferrari 1984, p. 37).

On the industrial side, the availability of funds and the justification of their expenditure on linkage programs are important constraints. Time and dollar commitments may be substantial and the potential payback highly speculative: "The stakes are high and so are the risks" (Prager and Omenn 1980, p. 380).

Organizational differences
As organizations, academia and industry differ. The emphasis in colleges and universities on educating students and conducting research relates to their basic roles of disseminating and extending knowledge. Industry's emphasis on commercialization and proprietary knowledge arises from its objectives of competitive edge and profitability. Moreover, for institutions of higher education, research productivity and quality of education are tied to freedom of inquiry and the open exchange of ideas, while industry's concern for financial viability and profit dictates setting priorities and timetables in line with corporate objectives (Prager and Omenn 1980, p. 380).

Because of different objectives, the time frame for expected results differs (Brodsky, Kaufman, and Tooker 1980, p. 7), and more emphasis is placed on interdisciplinary work in industry than normally found in academia (Sharp and Gumnick 1980, pp. 16–17). Values and attitudes diverge, with the inevitable stereotyping of academics as ivory tower theorists concerned with publications rather than with problems in the real world and of industrialists as overly directive, profit-hungry, and unconcerned with fundamental investigation. These attitudinal problems inhibit meaningful communication, impede cooperative efforts, and are "the most difficult barriers to overcome" (Prager and Omenn 1980, pp. 380–81).

The frequency of recommendations to improve communication and mutual understanding can be understood in the light of these differing objectives, values, and attitudes.

> *Industry cannot get instant knowledge any more than a professor can get instant experience. Industry must give its staff time to find out what is going on in academia and in professional societies, and professors must spend some of their time studying what is happening in industry* (Rahn and Segner 1976, p. 794).

Organizational rigidities

The key to successful research interactions is the effort of enterprising individuals, but such persons are often frustrated in developing a continuing relationship "not by the other party but by rigidities within their own organization" (National Science Foundation 1982, p. 30). Top-level commitment to cooperative ventures on both sides is crucial because of the flexibility needed to mount such programs and the freedom participants must have from other pressures on time and work (Brodsky, Kaufman, and Tooker 1980, p. 79).

A fundamental problem on the industrial side is that executives often fail to understand the impact of research on the technology base of their operations and the potential benefits of interaction with university faculty (Sharp and Gumnick 1980, pp. 15–17). Managers need to be convinced that research can pay off for their companies and that intersector collaboration is workable. On the academic side, top-level administrators need to adjust policies to encourage cooperative involvement and interdisciplinary research by faculty. Limiting recognition of academic accomplishment to highly specialized work in single disciplines is a major roadblock to cooperation with industry (Sharp and Gumnick 1980).

Organizational rigidities are also barriers to joint instructional programs. Many of the new degree-granting programs in the private sector and many in-house corporate training programs as well were developed only after the failure of earlier attempts to work out cooperative arrangements with colleges and universities (Cross 1981, p. 4). Disciplinary compartmentalization, rigid adherence to fixed time periods, isolation from practical realities, and lack of adaptation to individual student's needs or to new instructional technologies were among the major complaints from company representatives (pp. 4–5).

Awareness of these general barriers can provide a realistic assessment of the difficulties as well as the opportunities associated with various cooperative efforts. Additional insights can be gained by an analysis of successful and unsuccessful cases. Unsuccessful cooperative research projects, for example, lacked a continuing commitment from the company and included academics who promised more than they could deliver. But above all, "a communication gap resulting from a lack of time and effort put into building up a trust relationship between the two parties" was evident (Peters and Fusfeld 1983, p. 42).

Limiting recognition of academic accomplishment to . . . work in single disciplines is a major roadblock to cooperation with industry.

The problem in analyzing outcomes is that operational measures are seldom anticipated before cooperative projects begin. If it is indeed true that the level or extent of interaction affects results, it should be possible to measure the relationship between desired outcomes on the one hand and indicators of interaction on the other. The latter might include the frequency of interpersonal contacts, the time period covered by the project, the level of commitment in time, money, and manhours, or the number of people and organizations involved in various activities. To date, very little analysis of this kind has been focused on the various models of cooperation.

Dealing with Policy Issues

The higher education community is far from united on the merits of closer alliances with industry. To some academics, the statements of prominent institutional leaders advocating stronger links represent misguided measures to reap the benefits of corporate largess at the expense of academic freedom. Derek Bok, president of Harvard University, has described the evolution of the concepts of academic freedom, autonomy, and neutrality in the early years of this century as means of resisting interference in academic matters by powerful trustees, many of whom were wealthy industrialists (1982, pp. 5–7). The problem, as Bok acknowledges, is that these principles are much harder to adhere to now at the end of the century when colleges and universities are deeply enmeshed in public affairs.

The issues that arise in academic/industrial relations, then, must be addressed not only with the practical demands of commerce and government in mind but also with concern for the long-range interests and mission of the academy. Three such issues deserve special mention: (1) intellectual property rights; (2) nontraditional delivery of instruction; and (3) financial ties with the private sector.

Intellectual property rights

The disposition of patents and the publication of research results are sometimes the object of lengthy negotiations between academic and industrial participants in joint ventures. The university's interest is

> *. . . to assure that its patentable inventions will be fully and beneficially used, and that knowledge with a potential benefit to society at large will reach the public in a timely and useful fashion* (Giamatti 1982, p. 1280).

Accordingly, universities often prefer to grant nonexclusive licenses to make knowledge widely available, although in some cases, exclusive licenses may serve society better. In the communication of research results, any restrictions on publication should be avoided "save the most minor delay to enable a sponsor to apply for a patent or license" (Giamatti 1982, p. 1280). Moreover, any restrictions on free inquiry or oral communication of research results are totally unacceptable (Giamatti 1982).

The industrial viewpoint differs.

> *The exclusive dominion over the "property" resulting from funded research provides benefits that offset the competitive risks involved in spending on innovation while your competitor conserves his resources until he can spend them on imitation* (Kiley 1983, pp. 64–65).

Exclusive licenses provide companies an opportunity for a return on their investment and an incentive for additional expenditures "to pull the research results into the marketplace" (Kiley 1983, p. 65).

Actual agreements on these matters vary; they appear to be evolving. Some universities maintain strict policies and insist on reserving patent rights even in cooperative endeavors, while others emphasize basic guidelines with room for flexibility on specific projects (Brodsky, Kaufman, and Tooker 1980, pp. 31–32). An increasing number of institutions are developing their own patent management organizations. With some exceptions, companies generally feel comfortable with this approach, particularly if institutions provide exclusive licenses for a certain time period (National Science Foundation 1982, p. 25). Numerous universities are reviewing their policies on patents and licensing or have recently revised them to adapt to changing opportunities (Peters and Fusfeld 1983, p. 100).

When findings are published, corporations sometimes seek special guarantees concerning dissemination of information. The Monsanto–Washington University agreement for medical research on proteins and peptides, for example, has a secrecy clause. Technical developments may not be published without the company's approval and must remain secret until published. Faculty may also be required to sign statements committing them to confidentiality (Bouton 1983, p. 126). The typical period for prepublication review at 39 universities in one survey was from one to six months (Peters and Fusfeld 1983, p. 38). The agreement between Hoechst, A. G., and Massachusetts General Hospital at Harvard simply requires that papers be sent to the company 30 days before submission to journals (Bouton 1983, p. 126).

Corporate personnel do not appear to regard these issues as barriers to cooperation to the same extent that academics do, perhaps because they tend to view them as capable of resolution through negotiation (Cromie 1983, p. 252; Peters and Fusfeld 1983, p. 38). Although the orientation of academia and industry differs, it is often possible to arrive at a "reasonable compromise" (Jefferson 1982, p. 260).

Nontraditional delivery of instruction

Nontraditional study and continuing education programs have expanded in recent years in response to the phenomenal increase in the number of adults returning to school for educational experiences of all kinds. Surging adult enrollments and awareness of a still-larger potential market of adults have produced a large body of literature concerned with the needs and characteristics of adults as learners and with the challenge to traditional providers of education to adapt their programs and services.

Many of the recommended changes are relevant to the educational and training requirements of the private sector. Offering external study opportunities through unconventional scheduling or media, awarding credit for learning gained outside the academic setting, and joining employers in tailoring programs for workers all represent ways to bring education and the workplace closer together. While representatives of business and industry applaud these efforts, however, they often generate concern and resistance in academic quarters. Much of this resistance focuses on academic standards and on control of the curriculum.

One imaginative approach to academic/industrial cooperation illustrates the problem of nontraditional education (Cross 1981, p. 5). The John Wood Community College in Illinois works cooperatively with the Harris Corporation and with a private liberal arts college in a "common market" approach. The community college diagnoses community and student needs and acts as a "broker" for the delivery of instruction. The corporation offers technical instruction in broadcast electronics technology and provides physical facilities and sophisticated equipment. The liberal arts college contracts to provide general and liberal arts studies. The community college itself offers remedial education, social facilities, counseling, and administrative services.

These and similar arrangements may meet objections from many faculty on the grounds that they substitute the broker role for that of sole provider of instruction and give too strong a hand to industrial firms in determining curriculum. While the concerns are valid, the question should not be debated simply in terms of autonomy but in terms of the extent to which the arrangement contributes effectively to the education of students (Cross 1981). Perhaps the more fundamental point is that "learning or knowledge resides in the individual rather than in the courses offered by providers" (Cross 1978, p. 23).

Some writers have suggested that nontraditional methods like experiential learning assessment raise such basic issues about the educational process that faculty esist them because they are threatening (Meyer 1975, p. 15). On the other hand, advocates of various innovations need to take certain objective and legitimate factors into account. For example, in an empirical study of faculty attitudes toward external degree programs that discovered respondents tended to be quite receptive to the proposed innovation, the factors most strongly correlated with receptivity were the estimated feasibility in the professor's own field, the extent of agreement with the goal of greater access, and the desirability of various alternative methods of delivery (Johnson 1984, p. 493).

It is too simplistic to assume that educators are simply opposed to change (Craig and Evers 1981, pp. 42–43). Concerns about diminished standards and autonomy can be dealt with constructively only by bringing academics and employers together to discuss them directly. Adjustments in the thinking of both sectors are necessary, and developing strong working relationships is critical if both are to contribute effectively to the manpower needs of the nation (Lynton 1981, p. 4).

Financial ties with the private sector

As industrial and academic organizations move closer in new ventures, another emerging problem area is that of financial ties between them. The problem may occur at the individual level through consulting relationships or at the organizational level through investment. Financial connections arise in the realm of service through technology transfer and entrepreneurial development, but they pose serious problems for academic objectivity.

That academic freedom is essential for both scientific excellence and objectivity has been forcefully argued (Bok 1982, pp. 17–36). The influence of exciting opportunities to consult or render other services for external organizations may pose even greater dangers to scholarship than conventional attacks on academic freedom, however, causing professors to become more cautious and less able to maintain a detached viewpoint (p. 25).

One of the ways objectivity can be undermined is through a faculty member's own financial ties with private corporations. Such ties are "substantial" if a professor becomes a manager of a company in his or her area of research or acquires a significant share of stock in such a company (Giamatti 1982, p. 1279). The conflicts that then arise are threefold: diversion of time and energy from university work, conflicts regarding dissemination of knowledge, and ambiguities in the direction of research and relationships with one's colleagues and students. In such cases, it may be advisable to ask the faculty member to relinquish his or her academic appointment. As a result of such concerns, Yale, Harvard, the University of California, and other institutions are now requiring faculty members to disclose annually their connections with corporations and are developing guidelines to deal with cases of extensive involvement (Bouton 1983, pp. 151–52).

A somewhat parallel danger to objectivity can arise at the organizational level as well. Financial pressures have prompted colleges and universities to consider the possibility of increasing revenues by exploiting the products of academic research (Fusfeld 1981, pp. 4–5). It is a natural enough step from developing the patents an institution owns to considering ownership in the spin-off companies it spawns, especially when so much publicity accompanies the fortunes earned by new ventures in biotechnology and other areas.

Financially supporting firms started by an institution's own faculty has clear advantages, as well as dangers (National Governors Association 1983, p. 14). It enables the university to encourage technological advancement while obtaining equity in a potentially huge new source of revenue. But most observers to date agree that the dangers far outweigh the advantages. Institutional neutrality might be undermined, as financial considerations begin to influence the recruitment and treatment of faculty, the admission and opportunities available to graduate students, and the publication of research findings (Fusfeld 1981, p. 5). Harvard's decision in 1980 not to enter a commercial venture with some of its faculty was based on similar considerations—on the realization "that [the] pathway to riches would be marked by every kind of snare and pitfall" (Bok 1982, p. 160). Third-party mechanisms may be a much better approach (Fusfeld 1981, p. 6; National Governors Association 1983, p. 31).

The challenge in working with industry in research, teaching, and service, then, is to pursue new opportunities in ways that do not distort or limit freedom of inquiry, dilute high standards of instruction, or diminish the academy's objectivity and integrity.

> *These opportunities should not drive us toward arrangements . . . that abridge our principles. . . . We should negotiate appropriate arrangements, openly arrived at, that can further our mission* (Giamatti 1982, p. 1280).

Summary

This chapter has considerably expanded the horizon of academic/industrial relations, replacing the conceptual perspective based chiefly on the research paradigm by a broader framework embracing cooperative relations in human resource development, in research, and in entrepreneurial development. Specific mechanisms of cooperation in each of these areas can be categorized by three general levels of interaction. At the highest level of interaction, academic/industrial partnerships are the form of relationship most often recommended for effective contributions to technological development but least often discovered in present practice.

Balancing higher education's opportunities and obligations with its own mission and principles is a delicate matter.

> *Clearly, higher education's role is extremely important in the development of high technology industry. The question facing educational, political, and business leaders is specifically how higher education can best be part of economic development activities while maintaining its general mission for society* (Southern Regional Education Board 1983b, pp. 8–9).

CONCLUSION

Technology is of the earth, earthy; it is susceptible to pressure from industry and government departments; it is under an obligation to deliver the goods. . . . The attitude of universities toward technology is still ambiguous; until the ambiguity is resolved the universities will not have adapted themselves to one of the major consequences of the scientific revolution (Ashby 1958, p. 66).

Today's widespread interest in [academic/industrial] linkages . . . results from powerful pressures to strengthen the nation's technological capabilities.

This report looks at academic/industrial relationships from the viewpoint of economic development or, more specifically, of development through technological innovation. Today's widespread interest in linkages between higher education and industry results from powerful pressures to strengthen the nation's technological capabilities in the face of worldwide economic competition. Numerous states have launched ambitious plans for technological development designed in part to forge a high-technology connection between higher education and industry in the areas of research, manpower training, and technology transfer.

Cooperation in research and development typically involves the leading research universities and the large corporations that undertake extensive research and development themselves. The long tradition of relationships between American universities and corporations, marked by periodic ups and downs, has recently taken on a new intensity, especially in fields of very rapid technological change. Among the incentives for stronger research ties are the universities' needs to augment federal funding and industry's needs to cope with increasing competition and maintain access to science-based technology. Several models of cooperation in research are common, ranging from short-term contracts involving one institution and one company to cooperative research centers that may involve multiple institutions, companies, and purposes.

Cooperation in technology transfer is of interest to economic planners because of the view that advances in science and technology contribute to economic development only when they are used in the marketplace. Invention is but one part of the process of technological innovation. Several stages, considerable cost and risk, and countless factors affect its success. The circumstances of a particular community, an existing corporation, or a small business enterprise are all relevant to effective transfer, and they challenge academic institutions to work with other organizations in providing an array of services. In their most advanced form, the mechanisms used represent a cooperative approach to entrepreneurial development that goes far beyond transmittal of information.

Technological progress also depends on scientific and engineering manpower, training and retraining of the work force, and information on future occupational requirements. Collegiate and corporate organizations work together to address such needs through advisory structures, traditional and nontraditional programs, and other means. Thus, a conceptual framework for intersector relations must include the human resources dimension along with research activities and entrepreneurial services to encompass all of the substantive areas of interaction. Such a framework should also indicate that cooperative models vary in the extent of interaction involved, some being merely industry oriented, others collaborative in nature, and still others—however few in number—representing full partnerships. A total perspective further recognizes that serious barriers to cooperation will be encountered and that fundamental policy issues will arise, the first reflecting the difficulties of interorganizational effort and the second, the inevitable tensions between the role of service and the principles of academic freedom, autonomy, and objectivity.

Expectations and Constraints

Higher education and the corporate community share a common history of productive relationships in all of these major areas. Colleges and universities are already contributing significantly to the economic well-being of the nation, as they have been doing since the earliest years of the republic. Why, then, should academic/industrial linkages suddenly seem so vital to so many observers of the contemporary scene? Is there really any need to place this matter high on the agenda for planning and action?

The high-technology connection is a matter of considerable urgency for several reasons:

- The extraordinarily widespread concern throughout the nation for economic improvement stemming from decreased productivity, massive unemployment, and

major losses in market shares of manufactured goods to foreign competitors;
- The pervasive sense that the socioeconomic character of society is changing, with concomitant needs to reapply our intellectual and material resources in areas of greatest promise, such as advanced technology;
- The almost universal conviction that no sector of society is unaffected or can remain aloof from efforts to improve the economy and that in this regard public/private partnerships are imperative;
- The unprecedented energy and resources being committed to develop and launch comprehensive development strategies in almost every state of the union;
- The startling fact, in view of expressions of public disaffection with higher education in the recent past, that public officials are once more turning to academia to play a major role in what is probably the foremost concern of many communities, states, and regions.

The realization of what is expected of the academic community is even more sobering in light of the constraints with which institutions must deal. The most obvious limiting factor pertains to resources.

To spark discovery at the cutting edge of science requires sufficient numbers of highly competent research staff, well-equipped laboratories, up-to-date facilities, and adequate support services. To participate in the development of human resources through education and training requires a full complement of instructional staff, a reasonable ratio of faculty to students, and the kinds of salary scales and working conditions necessary to attract and hold the ablest professors. And to provide the additional services to corporations, small firms, and communities that will promote entrepreneurial development, colleges and universities must have the means to establish and staff service units with competent professionals. Nearly six out of ten faculty members feel their departments do not have sufficient resources to carry out their present missions (Watkins 1983, p. 19). Among those most concerned are faculty in business, science, and engineering—the very areas most likely to be called upon for development projects.

State planners, federal funding agencies, or public officials, including those most anxious for higher education to contribute to economic development, do not always recognize these needs, accounting for the fact that most college and university presidents regard inadequacies in basic support as the major obstacle to academic involvement with industry (Ferrari 1984, p. 14). It also accounts for the universal agreement that sustained levels of federal support are crucial to sustain higher education's research capabilities (Fusfeld 1983, p. 15).

In this connection, it is important to note that industrial support, even if dramatically increased, cannot take the place of funding from public sources for academic science. Even if corporate contributions were to double or even triple, they would still cover only a small portion of academic R&D (National Science Foundation 1982, p. 28). Or, to look at the matter another way, each 1 percent cut in federal research funding requires nearly a 25 percent increase in industrial support to maintain the present level of funding (McCoy, Krakower, and Makowski 1982, p. 349).

But other factors limit academic/industrial relationships as well. Our knowledge of innovation is limited. Technological innovation is a complex process that is not well understood. The same is true of our knowledge of organizational innovation, especially in the case of forging new structures that span organizational boundaries. Machiavelli's observation that "there is nothing more difficult to carry out, nor more doubtful of success, nor more dangerous to handle, than to initiate a new order of things" is still apt (1950, p. 21). "Creating a [new organization] is conceptually and action-wise as complex a task as can be undertaken . . ." (Sarason 1972, p. 21). Yet the usual tendency is to underestimate the difficulties and to fail to anticipate problems before they arise (p. 76).

Intersector endeavors are indeed difficult to develop and manage. It would help if the demanding nature of such efforts were recognized and appropriate steps taken at the outset. One important measure is to provide for the kind of leadership that can foster communication and trust and cope with the inevitable conflicts that arise when very different organizations attempt to work together.

All of these limitations and constraints are sufficient grounds to worry about promising more than can be delivered. But when economic development is tied to political agendas, it may be hard to avoid.

> *Modesty . . . is not the hallmark of political initiatives; consequently, one must beware the danger of overselling university-industry collaboration as an innovation "breakthrough"* (Baer 1980, p. iv).

Academic institutions, nevertheless, can at least temper their own public statements and perhaps support a stronger role in state development planning for their state coordinating agencies. State higher education boards are familiar with institutional constraints and with their needed resources as well; they can therefore exert a moderating influence if they have the opportunity. This matter will undoubtedly receive increasing attention in the future, as it is a weakness in many current state initiatives and because better coordination is necessary to use effectively the various types of postsecondary institutions in each state.

Strengthening Industrial Relationships

Ultimately, the role played by each institution will be determined by its own faculty, administration, and board of trustees. Not all institutions will wish to emulate Stanford and MIT or the many community and technical colleges whose participation with industry is heralded so often. At such institutions, it appears that "the philosophy of industrial collaboration has been fully integrated into the academic mission . . ." (General Accounting Office 1983, p. 48).

Many institutions, however, appear to have a strong interest in strengthening their industrial relationships. At most of the institutions in one survey, the presidents did not give very high marks to their present cooperative arrangements with industry, but they did rate the potential for increased cooperation during the 1980s as very high. Moreover, this positive interest was evident at all types of institutions in all parts of the country (Ferrari 1984, p. 20).

Colleges and universities have a number of options for improving relationships with industry, creating new positions—an "ombudsman" who can make it easier for outsiders to find the resources they need within the institution (Lee 1982, p. 131) or an "academic entrepreneur" who can effectively link academic personnel with corporate needs (Western Interstate Commission 1980, p. 35)—or new units—an "information, monitoring, and innovation center" to provide systematic information on the state of new technology (Bugliarello and Simon 1976, p. 79), a "regional services institute" where faculty and students would be involved in interdisciplinary research, teaching, and service (McGarrah 1981, p. 133), or a "professional practice plan" to channel faculty consulting or training activities through the institution (Linnell 1982b, pp. 130–32; Lynton 1981, p. 13).

Within every institution, many current activities relate in some way to industry. Within comprehensive universities, they may be found in all three areas of cooperation: research, human resource development, and entrepreneurial development. A good way to begin might be to review the programs in each of these areas and determine how effectively they are working and how they might be improved. Such deliberations might become part of the institution's strategic planning process, with periodic monitoring of cooperative activities and outcomes in relation to changing needs (Kyle 1981, p. 100).

Some of the activities an institution might consider undertaking are listed in table 4. These activities constitute a kind of "industrial relations audit." They may or may not lead to recommendations for new positions, new units, or revised policies and programs. Certainly they should not be undertaken without the participation of faculty and of industrial and community representatives as well.

Institutional initiatives of this kind do *not* mean that all academic endeavors should be judged in economic terms. A commitment to contribute to economic development need *not* be incompatible with academic values, so long as it is implemented on terms defined by the academy. In times of transition, it is necessary to "completely rethink what it is that you are doing" (Naisbitt 1982, p. 86). Clearly it is time for higher education to do just that with regard to its relations with industry.

TABLE 4
RECOMMENDATIONS FOR AN INDUSTRIAL RELATIONS AUDIT

Inventory and evaluate cooperative activities already in place in the areas of research, human resources, and entrepreneurial services. Ask representatives of business and industry to assist.

Survey the interests and expertise of faculty related to industrial and technological matters; do not assume that this recommendation applies only to science, engineering, technology, or business faculty.

Work with corporations, small business associations, local community growth organizations, and government agencies to assess the private sector's needs in the region. Identify gaps in services, rank unmet needs, and consider specific actions to enhance responsiveness.

Review and evaluate departmental, collegiate, and institutional policies affecting faculty participation (time, reporting, recognition, rewards), the use of facilities and other resources, and the treatment of income and expenditures.

Review institutional policies regarding patents and licensing to ascertain whether they balance the interests of industrial supporters, the institution itself, and the community at large.

Review institutional practices in the area of nontraditional programs and services to assess the extent to which adult learners' needs are met and academic standards are maintained.

Review policies pertaining to consulting to determine whether they encourage appropriate use of faculty expertise without detracting from other academic responsibilities and without diminishing opportunities for the institution as a whole to respond to industry.

Consider mechanisms for overall improvement of linkages, such as special liaisons, service units or practice plans, industrial representatives boards, and so on.

Survey local, state, and federal programs and funding opportunities to identify those that match local or regional needs. Assign specific individuals to gather detailed information on relevant areas and to develop proposals for funding.

Invite faculty in nontechnical disciplines to develop courses, seminars, or series dealing with technology and its implications for society and the quality of life.

REFERENCES

The ERIC Clearinghouse on Higher Education abstracts and indexes the current literature on higher education for the National Institute of Education's monthly bibliographic journal *Resources in Education*. Most of these publications are available through the ERIC Document Reproduction Service (EDRS). For publications cited in this bibliography that are available from EDRS, ordering number and price are included. Readers who wish to order a publication should write to the ERIC Document Reproduction Service, P.O. Box 190, Arlington, Virginia 22210. When ordering, please specify the document number. Documents are available as noted in microfiche (MF) and paper copy (PC). Because prices are subject to change, it is advisable to check the latest issue of *Resources in Education* for current cost based on the number of pages in the publication.

Abernathy, William J., and Rosenbloom, Richard S. 1982. "The Institutional Climate for Innovation in Industry: The Role of Management Attitudes and Practices." In *The Five-Year Outlook on Science and Technology, 1981*, prepared by the American Association for the Advancement of Science. Washington, D.C.: National Science Foundation.

American Association of Community and Junior Colleges, and the Association of Community College Trustees. 1983. *Putting America Back to Work: Report of the Phase I Kellogg Leadership Initiative*. Washington, D.C.: Author.

Ashby, Sir Eric. 1958. *Technology and the Academics*. London: MacMillan and Company, Ltd.

Baer, Walter S. 1977. "University Relationships with Other R&D Performers." Santa Monica, Calif.: The Rand Corporation. ED 144 468. 77 pp. MF–$1.17; PC–$9.37.

——— 1980. "Strengthening University-Industry Interaction." Santa Monica, Calif.: The Rand Corporation. ED 190 033. 34 pp. MF–$1.17; PC not available EDRS.

Baker, William O. 1983. "Organizing Knowledge for Action." In *Partners in the Research Enterprise: University-Corporate Relations in Science and Technology*, edited by Thomas W. Langfitt, Sheldon Hackney, Alfred P. Fishman, and Albert V. Gowacky. Philadelphia: University of Pennsylvania.

Battenburg, Joseph R. 1980. "Forging Links between Industry and the Academic World." *Journal of the Society of Research Administrators* 12 (3): 5–12.

Birch, David L. 1979. *The Job Generation Process*. Report of the MIT Program on Neighborhood and Regional Change to the Economic Development Administration, U. S. Department of Commerce. Cambridge: Massachusetts Institute of Technology.

Birchfield, Jerry. 1982. "High Technology Centers of Excellence." Paper presented at the Georgia Advanced Technology

Strategy Conference, October, Georgia Institute of Technology, Atlanta, Georgia.

Birr, Kendall A. 1966. "Science in American Industry." In *Science and Society in the United States*, edited by David D. Van Tassel and Michael G. Hall. Homewood, Ill.: Dorsey Press.

Bok, Derek. 1982. *Beyond the Ivory Tower: Social Responsibilities of the Modern University*. Cambridge: Harvard University Press.

Bouton, Katherine. 11 September 1983. "Academic Research and Big Business: A Delicate Balance." *The New York Times Magazine:* 62–63 + .

Brodsky, Neal H.; Kaufman, Harold G.; and Tooker, John D. 1980. *University/Industry Cooperation: A Preliminary Analysis of Existing Mechanisms and Their Relationship to the Innovation Process*. New York: Center for Science and Technology Policy, Graduate School of Public Administration, New York University.

Brophy, David J. 1974. *Finance, Entrepreneurship, and Economic Development*. Ann Arbor: Industrial Development Division, Institute of Science and Technology, University of Michigan.

Bruce, James D., and Tamaribuchi, Kay. 1981. "MIT's Industrial Liaison Program." *Journal of the Society of Research Administrators* 12 (3): 13–16.

Bugliarello, George, and Simon, Harold A. 1976. *Technology, the University, and the Community: A Study of the Regional Role of Engineering Colleges*. Elmsford, N.Y.: Pergamon Press.

Cameron, S. W. 1980. "Career Planning and Learning Assessment in Colleges and Universities in Ohio." Mimeographed. Columbus: Ohio Board of Regents.

Carter, Luther J. June 1978. "Research Triangle Park Succeeds beyond Its Promoters' Expectations." *Science* 200 (4349): 1469–70.

Colton, Robert M. 1978. "Technological Innovation through Entrepreneurship." *Engineering Education* 69 (2): 193–95.

Commission on Nontraditional Study. 1973. *Diversity by Design*. San Francisco: Jossey-Bass.

Committee for Economic Development. 1980. *Stimulating Technological Progress*. New York: Committee for Economic Development.

Cooper, Martin J. 1979. "Universities and the Private Sector: Opportunities for Mutual Gain in the Decade Ahead." *Journal of the Society of Research Administrators* 3 (3): 27–31.

Craig, Robert L., and Evers, Christine J. 1981. "Employers as Educators: The 'Shadow Education System.' " In *Business and Higher Education: Toward New Alliances,* edited by Gerald G. Gold. New Directions for Experimental Learning No. 13. San Francisco: Jossey-Bass.

Cromie, William. 1983. "University-Industry Cooperation in Microelectronics and Computers." In *University-Industry Research Relationships*. Washington, D.C.: National Science Foundation. ED 243 343. 298 pp. MF–$1.17; PC–$24.14.

Cross, K. Patricia. 1978. *The Missing Link: Connecting Adult Learners to Learning Resources*. New York: College Entrance Examination Board. ED 163 177. 87 pp. MF–$1.17; PC not available EDRS.

———. 1981. "New Frontiers for Higher Education: Business and the Professions." In *Partnerships with Business and the Professors*. Current Issues in Higher Education No. 3. Washington, D.C.: American Association for Higher Education. ED 213 325. 27 pp. MF–$1.17; PC not available EDRS.

Crosson, Patricia H. 1983. *Public Service in Higher Education: Practices and Priorities*. ASHE-ERIC Higher Education Research Report No. 7. Washington, D.C.: Association for the Study of Higher Education.

Culliton, Barbara J. May 1982. "The Academic-Industrial Complex." *Science* 216: 960–62. ED 239 569. 140 pp. MF–$1.17: PC–$12.87.

David, Edward E., Jr. 1983. "High Tech and the Economy: 'What Does It All Mean?' " *Research Management* 26 (5): 27–30.

DeMeester, Lynn A. 1981. "Incentives for Learning and Innovation." In *Business and Higher Education toward New Alliances,* edited by Gerald G. Gold. New Directions for Experiential Learning No. 13. San Francisco: Jossey-Bass.

Dillon, Kristine E. 1982. "Economics of the Academic Profession: A Perspective on Total Professional Earnings." In *Dollars and Scholars,* edited by Robert H. Linnell. Los Angeles: University of Southern California Press.

Drucker, Peter F. 1984. "Our Entrepreneurial Economy." *Harvard Business Review* 62 (1): 59–64.

Dupree, A. Hunter. 1957. *Science and the Federal Government: A History of Policies and Activities to 1940*. New York: Harper & Row.

Edling, Walter. 1982. "The Effects of High Technology on Two-Year College Programs." *The Associate*. Columbus, Ohio: Ohio Community and Technical College Association.

Fernelius, W. Conrad, and Waldo, Willis H. 1980. "Role of Basic Research in Industrial Innovation." *Research Management* 23 (4): 36–40.

Ferrari, Michael A. 1984. "National Study of Cooperation between Higher Education and Industry." Paper read at the annual meeting of the Ohio Academy of Science, April, Case Western Reserve University, Cleveland, Ohio.

Finkbeiner, Howard F. 1969. "Formation and Administration of Selected Interdisciplinary Research Units in a Large State University." Ph.D. dissertation, University of Michigan.

Finniston, Sir Montague. 1979. *Engineering Our Future: Report of the Committee of Inquiry into the Engineering Profession*. London: Office of the Secretary of State for Industry.

Fusfeld, Herbert I. 1981. "The Role of University Research in New Business Development." Mimeographed. New York: Center for Science and Technology Policy, New York University.

———. 1983. "Overview of University-Industry Research Interactions." In *Partnerships in the Research Enterprise: University-Corporate Relations in Science and Technology*, edited by Thomas W. Langfitt and associates. Philadelphia: University of Pennsylvania Press.

Geils, John W. 1983. "The Faculty Shortage: The 1982 Survey." *Engineering Education* 74 (1): 47–53.

General Accounting Office. 1983. *The Federal Role in Fostering University-Industry Cooperation*. Washington D.C.: U.S. General Accounting Office.

Georgia Office of Planning and Budget. 1982. "Georgia's Economic Future: The Technology Adjustment." Mimeographed. Atlanta: Office of Planning and Budget, State of Georgia.

Giamatti, A. Bartlett. 1982. "The University, Industry, and Cooperative Research." *Science* 218 (24): 1278–80.

Gold, Bela; Rosegger, Gerhard; and Boylan, Myles G., Jr. 1980. *Evaluating Technological Innovations*. Lexington, Mass.: D.C. Heath & Co.

Gold, Gerald G. 1981. "Toward Business–Higher Education Alliances." In *Business and Higher Education: Toward New Alliances*, edited by Gerald G. Gold. New Directions for Experiential Learning No. 13. San Francisco: Jossey-Bass.

Grad, Marcia, and Shapero, Albert. 1981. "Federal and State Policies for Entrepreneurship Education." Working Paper 81–84. Columbus, Ohio: College of Administrative Science, Ohio State University.

Hansen, Arthur G. 1983. "The Industrial-Educational Partnership: The Promise and the Problems." *Technological Horizons in Education* 11 (2): 113–16.

Harbison, Frederick, and Myers, Charles M. 1964. *Education, Manpower, and Economic Growth: Strategies of Human Resource Development*. New York: McGraw-Hill.

Hayes, Robert H., and Abernathy, William J. 1980. "Managing Our Way to Economic Decline." *Harvard Business Review* 58 (4): 67–77.

Healey, Frank H. 1978. "Industry's Needs for Basic Research." *Research Management* 21 (6): 12–16.

Heffernan, James M.; Macy, Francis U.; and Vickers, Donn F. 1976. *Educational Brokering: A New Service for Adult Learners*. Syracuse: National Center for Educational Brokering. ED 136 833. 93 pp. MF–$1.17; PC not available EDRS.

Helms, Clyde W., Jr. 1981. "Occupational Forecasting: An Emerging Management Requirement." Paper presented at the Twelfth National Conference on Human Resource Management Systems, Arlington, Virginia.

Hewlett, William, et al. 1982. "Research in Industry." In *Outlook for Science and Technology: The Next Five Years*, compiled by the National Research Council. San Francisco: W. H. Freeman & Co.

Hodges, Wayne. 1982. "High Technology Overview." Mimeographed. Paper read at the Georgia Advanced Technology Strategy Conference, October, Georgia Institute of Technology, Atlanta.

Holloman, J. Herbert. 1974. *Technical Change and American Enterprise*. Washington, D.C.: National Planning Association.

Holtzman, Steven. 1983. "The Thomas Alva Edison Partnership Program: Background Briefing Paper." Mimeographed. Columbus, Ohio: Department of Development, State of Ohio.

Honan, James P. 1982. "Corporate Education: Threat or Opportunity?" *AAHE Bulletin* 34 (7): 7–9.

Houle, Cyril O. 1974. *The External Degree*. San Francisco: Jossey-Bass.

Jamieson, David W., and Warren, James R. 1980. "Forces and Trends Affecting the Future: 1980–1990." Paper delivered at the national conference of the American Society for Training and Development, April, Anaheim, California.

Jefferson, Edward G. 1982. "The Impact of Polymer Science." *Polymer News* 8 (9): 258–62.

Johnson, Lynn G., ed. 1980. *Assessing the Needs of Adult Learners: Methods and Models*. Columbus, Ohio: Ohio Board of Regents. ED 199 387. 78 pp. MF–$1.17; PC not available EDRS.

———. July 1984. "Faculty Receptivity to an Innovation: A Study of Attitudes toward External Degree Programs." *Journal of Higher Education* 55 (4): 481–99.

Johnson, Wilson. 1978. "Meeting Small Business Needs through the SBDC." *AACSB Bulletin* 14 (3): 11–12.

Joint Economic Committee, United States Congress. 1982. *Location of High Technology Firms and Regional Economic Development*. Staff report prepared for the Subcommittee on Monetary and Fiscal Policy, Joint Economic Committee. Washington, D.C.: U.S. Government Printing Office.

Kerr, Clark. 1972. *The Uses of the University*. Cambridge: Harvard University Press.

Kiley, Thomas D. 1983. "Licensing Revenue for Universities: Impediments and Possibilities." In *Partners in the Research Enterprise*, edited by Thomas W. Langfitt and associates. Philadelphia: University of Pennsylvania Press.

Kyle, Regina M. J. 1981. "Partners in Economic Development." In *Business and Higher Education: Toward New Alliances*, edited by Gerald G. Gold. New Directions for Experimental Learning No. 13. San Francisco: Jossey-Bass.

Lamont, Lawrence M. 1971. *Technology Transfer, Innovation, and Marketing in Science-Oriented Spin-Off Firms*. Ann Arbor: Industrial Development Division, Institute of Science and Technology, University of Michigan.

Lee, Charles A. 1982. "University-Related Research Parks: A Michigan Case Study with Selected Comparisons." Ph.D. dissertation, University of Michigan.

Levy, Lawrence. 1977. "National Science and Technology Policy—Needed: Institutional Breakthroughs." *Research Management* 20 (1): 21–24.

Libsch, Joseph F. 1976. "Industrial/University R&D—The Role of the Small, High Technology University." *Research Management* 19 (3): 28–31.

Linnell, Robert H. 1982a. "Professional Activities for Additional Income: Benefits and Problems." In *Dollars and Scholars*, edited by Robert H. Linnell. Los Angeles: University of Southern California Press.

———. 1982b. "The University's Future: Summary Recommendations and Conclusions." In *Dollars and Scholars*, edited by Robert H. Linnell. Los Angeles: University of Southern California Press.

Long, James P., and Warmbrod, Catherine P. 1982. *Preparing for High Technology: A Guide for Community Colleges*. Prepared for the National Postsecondary Alliance. Columbus, Ohio: The National Center for Research in Vocational Education. ED 216 169. 23 pp. MF–$1.17; PC–$3.74.

Low, George M. 1983. "The Organization of Industrial Relationships in Universities." In *Partners in the Research Enterprise: University-Corporate Relations in Science and Technology*, edited by Thomas W. Langfitt and associates. Philadelphia: University of Pennsylvania Press.

Lynton, Ernest A. 1981. "A Role for Colleges in Corporate Training and Development." In *Partnerships with Business and the Professions*. Current Issues in Higher Education No. 3. Washington, D.C.: American Association for Higher Education. ED 213 325. 27 pp. MF–$1.17; PC not available EDRS.

McCoy, Marilyn; Krakower, J.; and Makowski, D. 1982. "Financing at the Leading 100 Research Universities: A Study of Financial Dependency, Concentration, and Related Institutional Characteristics." *Research in Higher Education* 16 (4): 323–53.

McGarrah, Robert E. 1981. "Expanding Higher Education's Role in New England's Economic Development." In *Business and Academia: Partners in New England's Economic Renewal*, edited by John C. Hoy and Melvin H. Bernstein. Hanover, N.H.: University Press of New England.

Machiavelli, Niccolo. 1950. *The Prince*. New York: Modern Library.

Magarrell, Jack. February 1984. "Governors Budget More Money for Colleges to Spur Economy, Make Up for Past Reductions." *Chronicle of Higher Education* 27 (24):1+.

Maidique, Modesto, and Hayes, Robert H. 1983. "The Art of High Technology Management." Working paper. Boston: Division of Research, Graduate School of Business Administration, Harvard University.

Mansfield, Edwin. 1968. *Industrial Research and Technological Innovation: An Econometric Analysis*. New York: W. W. Norton & Co.

Meyer, Peter. 1975. *Awarding College Credit for Noncollege Learning*. San Francisco: Jossey-Bass.

Millett, John D. 1968. *Decision Making and Administration in Higher Education*. Kent, Ohio: Kent State University.

Molitor, Graham T. 1981. "The Information Society: The Path to Post-Industrial Growth." *The Futurist* 15 (2): 23–27.

Naisbitt, John. 1982. *Megatrends: Ten New Directions Transforming Our Lives*. New York: Warner Books.

National Academy of Sciences. 1978. *Technology, Trade, and the U.S. Economy*. Washington, D.C.: National Academy of Sciences.

National Commission on Excellence in Education. 4 May 1983. "A Nation at Risk: The Imperative for Educational Reform." *Chronicle of Higher Education* 26 (10): 11–16.

National Commission on Research. 1980. *Industry and the Uni-*

versities: Developing Cooperative Research Relationships in the National Interest. Washington, D.C.: NCR. ED 201 217. 49 pp. MF–$1.17; PC–$5.49.

National Governors Association. 1983. *State Initiatives in Technological Innovation: Preliminary Report of Survey Findings*. Washington, D.C.: National Governors Association.

National Science Foundation. 1980. *Employment Patterns of Academic Scientists and Engineers*. Special report. Washington, D.C.: National Science Foundation. ED 195 189. 22 pp. MF–$1.17; PC–$3.74.

———. 1982. *University-Industry Research Relationships: Myths, Realities, and Potentials*. Fourteenth annual report of the National Science Board. Washington, D.C.: National Science Foundation. ED 230 115. 39 pp. MF–$1.17; PC–$5.49.

Office of Technology Assessment. 1983. *Technology, Innovation, and Regional Economic Development*. Washington, D.C.: Office of Technology Assessment. ED 235 344. 81 pp. MF–$1.17; PC–$9.37.

Ohio Department of Development. 1983. "Thomas Alva Edison Partnership Program: Advanced Technology Application Centers Program." General Guidelines and Application Instructions. Mimeographed. Columbus, Ohio: Ohio Department of Development.

Omenn, Gilbert S. 1983. "University-Corporate Relations in Science and Technology: An Analysis of Specific Models." In *Partners in the Research Enterprise: University-Corporate Relations in Science and Technology*, edited by Thomas W. Langfitt and associates. Philadelphia: University of Pennsylvania Press.

Organization for Economic Cooperation and Development. 1982. *The University and the Community*. Paris: Organization for Economic Cooperation and Development. ED 224 353. 157 pp. MF–$1.17; PC not available EDRS.

Park, J. C. 1983. "Entrepreneurial Corporations: A Managerial Assessment." *Journal of Small Business Management* 21 (4): 38–43.

Patterson, Franklin. 1974. *Colleges in Consort*. San Francisco: Jossey-Bass.

Pelz, Donald C., and Andrews, Frank M. 1966. *Scientists in Organizations: Productive Climates for Research and Development*. Ann Arbor, Mich.: Institute for Social Research, University of Michigan.

Pennsylvania Department of Commerce. 1983. "Pennsylvania's Advanced Technology Initiatives: The Ben Franklin Partnership Program." Mimeographed. Harrisburg: Pa. Dept. of Commerce.

Peters, Lois S., and Fusfeld, Herbert I. 1983. "Current U.S. University/Industry Research Connections." In *University-Industry Research Relationships*. Washington, D.C.: National Science Foundation. ED 243 343. 298 pp. MF–$1.17; PC–$24.14.

Peters, Thomas J., and Waterman, Robert H., Jr. 1982. *In Search of Excellence: Lessons from America's Best-Run Companies*. New York: Harper & Row.

Peterson, Iver. 24 October 1982. "Is Talk of High Tech Jobs More Political than Real?" *The New York Times*.

Peterson, Richard E. 1979. "Present Sources of Education and Training." In *Lifelong Learning in America*, edited by Richard E. Peterson and associates. San Francisco: Jossey-Bass.

Phalon, Richard. 1983. "University as Venture Capitalist." *Forbes* 132 (51): 82–93.

Pollack, Andrew. 25 March 1984. "Job Outlook for the '80s Is Generally Optimistic." *The New York Times*.

Powell, Reed M. 1978. "The Small Business Development Center Program." *AACSB Bulletin* 14 (3): 13–22.

Prager, Denis J. 1983. "Institutional Change: Impact on Science and Technology in the 80s." *Journal of the Society of Research Administrators* 14 (4): 5–10.

Prager, Denis, and Omenn, Gilbert S. 1980. "Research, Innovation, and University-Industry Linkages." *Science* 207 (4429): 379–84.

Pratt, Stanley. 1982. "The Role of Venture Capital in Georgia's High Technology Future." Mimeographed. Paper presented at the Georgia Advanced Technology Strategy Conference, October, Georgia Institute of Technology, Atlanta.

Rahn, H. W., and Segner, E. P., Jr. 1976. "Technical Research: Pathways for Improving Interaction between Academia and Industry." *Engineering Education* 66 (8): 794–853.

Rettig, Richard A. 1982. "Applying Science and Technology to Public Purposes: A Synthesis." In *The Five-Year Outlook on Science and Technology, 1981*, prepared by the American Association for the Advancement of Science. Washington, D.C.: National Science Foundation.

Roberts, Edward B. 1968. "A Basic Study of Innovators: How to Keep and Capitalize on Their Talents." *Research Management* 11 (4): 249–66.

Roberts, Edward B., and Peters, D. H. 1981. "Commercial Innovation from University Faculty." *Research Policy* 10: 108–26.

Ross, Steven S. 1983. "Incubating Industries on the Campus." *Graduating Engineer* 5 (1): 57–60.

Roy, Rustum. 1972. "University-Industry Interaction Patterns." *Science* 178: 955–59.

Rumberger, Russell W. 1983. "The Job Market for College Graduates, 1960–1990." Research report. Stanford, Calif.: Institute for Research on Educational Finance and Governance, Center for Educational Research, Stanford University. ED 229 880. 42 pp. MF–$1.17; PC–$5.49.

Sarason, Seymour B. 1972. *The Creation of Settings and the Future Societies*. San Francisco: Jossey-Bass.

Seitz, Frederick, et al. 1982. "Prospects for New Technologies." In *Outlook for Science and Technology: The Next Five Years*, compiled by National Research Council. San Francisco: W. H. Freeman & Co.

Seitz, Frederick, and Handler, Philip. 1982. "Observations." In *Outlook for Science and Technology: The Next Five Years*, compiled by National Research Council. San Francisco: W. H. Freeman & Co.

Shapero, Albert. 1979. "University-Industry Interactions: Recurring Expectations, Unwarranted Assumptions, and Feasible Policies." Mimeographed. Columbus, Ohio: Ohio State University.

———. 1982a. "Developing a High Tech Complex through Small Company Formations." *Survey of Business* 18 (1): 17–20.

———. 1982b. "Inventors and Entrepreneurs: Their Roles in Innovation." Working paper 82-35. Columbus, Ohio: College of Administrative Science, Ohio State University.

Sharp, James M., and Gumnick, James L. 1980. "University-Industry Connections—The GURC Example." *Journal of the Society of Research Administrators* 12 (2): 15–21.

Sirbu, Marvin A.; Treitel, R.; Yorz, W.; and Roberts, E. 1976. *The Formation of Technology-Oriented Complexes: Lessons from North American and European Experience*. Cambridge: Center for Policy Alternatives, Massachusetts Institute of Technology.

Smith, Bruce L. R., and Karlesky, Joseph J. 1977. *The Universities in the Nation's Research Effort*. New Rochelle, N.Y.: Change Magazine Press.

Sorrows, Howard E. 1983. "Future University-Industry Relations." *Journal of the Society of Research Administrators* 14 (3): 5–8.

Sosdian, Carol P., and Sharp, Laure M. 1977. *Guide to Undergraduate External Degree Programs in the United States*. Washington, D.C.: National Institute of Education. ED 155 981. 70 pp. MF–$1.17; PC–$7.24.

Southern Regional Education Board. 1983a. "Sites for High Technology Activities." Mimeographed. Atlanta: SREB.

———. 1983b. "Universities and High Technology Development." Mimeographed. Atlanta: SREB.

Star, Alvin D., and Narayana, Chem L. 1983. "Do We Really Know the Number of Small Business Starts?" *Journal of Small Business Management* 21 (4): 44–48.

Stever, Guyford H. 1972. "New Dimensions of Research Cooperation." *Research Management* 15 (3): 23–29.

Swalin, R. A. 1976. "Industry/University R&D: Improving Interaction between the University and the Technical Community." *Research Management* 19 (3): 25–27.

Thackray, Arnold. 1983. "University-Industry Connections and Chemical Research: An Historical Perspective." In *University-Industry Research Relationships*. Washington, D.C.: National Science Foundation. ED 243 343. 298 pp. MF–$1.17; PC–$24.14.

Thomas, Rose. 1983. "The New Corporate Campus: Universities and Private Industry Become Partners and Build Together." *Building Design and Construction* 24 (8): 74–79.

Tornatzky, Louis G.; Eveland, J. D.; Boylan, M. G.; Hetzner, W. A.; Johnson, E. C.; Roitman, D.; and Schneider, J. 1983. *The Process of Technological Innovation: Reviewing the Literature*. Washington, D.C.: National Science Foundation. ED 233 697. 284 pp. MF–$1.17; PC–$24.14.

Tornatzky, Louis G.; Hetzner, W. A.; Eveland, J. D.; Schwarzkopf, A.; and Colton, R. M. 1982. *University-Industry Cooperative Research Centers: A Practice Manual*. Washington, D.C.: Innovation Processes Research Section, Division of Industrial Science and Technological Innovation, National Science Foundation.

Tver, David F., comp. 1974. *The Gulf Publishing Company Dictionary of Business and Science*. Houston: The Gulf Publishing Co.

Urban Institute. 1983. *Directory of Incentives for Business Investment and Development in the United States: A State-by-State Guide*. Washington, D.C.: The Urban Institute.

Useem, Elizabeth. 1981. "Education and High Technology Industry: The Case of Silicon Valley. Summary of Research Findings." Boston: Northeastern University. ED 222 107. 34 pp. MF–$1.17; PC–$5.49.

———. 1982. "Education in a High Technology World: The Case of Route 128." Boston: Northeastern University. ED 222 108. 83 pp. MF–$1.17; PC–$9.37.

Valley, John R. 1979. "Local Programs: Innovations and Problems." In *Lifelong Learning in America,* edited by Richard E. Peterson and associates. San Francisco: Jossey-Bass.

Venture Capital Journal. August 1983. "Special Report—

Universities Emerge as an Important Catalyst in the New Business Development Process" : 7–12.

Votruba, J. C. 1978. "Faculty Rewards for University Outreach: An Integrated Approach." *Journal of Higher Education* 49 (6): 639–48.

Warmbrod, Catherine P.; Persavich, Jon J.; and L'Angelle, David. 1981. *Sharing Resources: Postsecondary Education and Industry Cooperation*. Columbus, Ohio: National Center for Research in Vocational Education, Ohio State University. ED 204 532. 150 pp. MF–$1.17; PC–$12.87.

Watkins, Beverly T. 1983. "Concern over Departments' Resources Found Widespread among Professors." *Chronicle of Higher Education* 27 (13): 19–20.

Wayne, Leslie. 25 March 1984. "A Pioneer Spirit Sweeps Business." *The New York Times*.

Weiner, Charles. 1966. "Science and Higher Education." In *Science and Society in the United States*, edited by David D. Van Tassel and Michael G. Hall. Homewood, Ill.: Dorsey Press.

Western Interstate Commission for Higher Education. 1980. *Higher Education and Economic Development in the West: Report of a Regional Conference*. Boulder, Colo.: Western Interstate Commission for Higher Education. ED 213 373. 61 pp. MF–$1.17; PC–$7.24.

White, Philip C. 1973. "Better Industry-University-Government Cooperation: Why and How." *Research Management* 16 (1): 10–15.

White, Philip C., and Wallin, C. C. September 1974. "The Role of Universities in Industrial Research: What Industry Needs from Academia." *Research Management* 17: 29–32.

Wilson, James W. No date. "Excerpts from the Commentary." Comments on *Cooperative Education—A National Assessment*. Pamphlet. Boston: National Commission for Cooperative Education.

Wilson, Joann. 1981. *A Study of the Relationship between Postsecondary Education and Economic Development in Selected States*. Phoenix: Arizona Office of Economic Planning and Development and Arizona Commission for Postsecondary Education. ED 215 649. 84 pp. MF–$1.17; PC–$9.37.

Wilson, Logan. 1979. *American Academics Then and Now*. New York: Oxford University Press.

Wood, James M. 1983. "The New Wave in American Success." *Cleveland Magazine* 12 (5): 58–63.

ASHE-ERIC HIGHER EDUCATION RESEARCH REPORTS

Starting in 1983, the Association for the Study of Higher Education assumed cosponsorship of the Higher Education Research Reports with the ERIC Clearinghouse on Higher Education. For the previous 11 years, ERIC and the American Association for Higher Education prepared and published the reports.

Each report is the definitive analysis of a tough higher education problem, based on a thorough research of pertinent literature and institutional experiences. Report topics, identified by a national survey, are written by noted practitioners and scholars with prepublication manuscript reviews by experts.

Ten monographs in the ASHE-ERIC Higher Education Research Report series are published each year, available individually or by subscription. Subscription to 10 issues is $55 regular; $40 for members of AERA, AAHE, and AIR; $35 for members of ASHE. (Add $7.50 outside U.S.)

Prices for single copies, including 4th class postage and handling, are $7.50 regular and $6.00 for members of AERA, AAHE, AIR, and ASHE. If faster 1st class postage is desired for U.S. and Canadian orders, for each publication ordered add $.75; for overseas, add $4.50. For VISA and MasterCard payments, give card number, expiration date, and signature. Orders under $25 must be prepaid. Bulk discounts are available on orders of 10 or more of a single title. Order from the Publications Department, Association for the Study of Higher Education, One Dupont Circle, Suite 630, Washington, D.C. 20036, (202) 296-2597. Write for a complete list of Higher Education Research Reports and other ASHE and ERIC publications.

1981 Higher Education Research Reports

1. Minority Access to Higher Education
   Jean L. Preer
2. Institutional Advancement Strategies in Hard Times
   Michael D. Richards and Gerald Sherratt
3. Functional Literacy in the College Setting
   Richard C. Richardson, Jr., Kathryn J. Martens, and Elizabeth C. Fisk
4. Indices of Quality in the Undergraduate Experience
   George D. Kuh
5. Marketing in Higher Education
   Stanley M. Grabowski
6. Computer Literacy in Higher Education
   Francis E. Masat
7. Financial Analysis for Academic Units
   Donald L. Walters

8. Assessing the Impact of Faculty Collective Bargaining
   J. Victor Baldridge, Frank R. Kemerer, and Associates
9. Strategic Planning, Management, and Decision Making
   Robert G. Cope
10. Organizational Communication in Higher Education
    Robert D. Gratz and Philip J. Salem

1982 Higher Education Research Reports

1. Rating College Teaching: Criterion Studies of Student Evaluation-of-Instruction Instruments
   Sidney E. Benton
2. Faculty Evaluation: The Use of Explicit Criteria for Promotion, Retention, and Tenure
   Neal Whitman and Elaine Weiss
3. The Enrollment Crisis: Factors, Actors, and Impacts
   J. Victor Baldridge, Frank R. Kemerer, and Kenneth C. Green
4. Improving Instruction: Issues and Alternatives for Higher Education
   Charles C. Cole, Jr.
5. Planning for Program Discontinuance: From Default to Design
   Gerlinda S. Melchiori
6. State Planning, Budgeting, and Accountability: Approaches for Higher Education
   Carol E. Floyd
7. The Process of Change in Higher Education Institutions
   Robert C. Nordvall
8. Information Systems and Technological Decisions: A Guide for Non-Technical Administrators
   Robert L. Bailey
9. Government Support for Minority Participation in Higher Education
   Kenneth C. Green
10. The Department Chair: Professional Development and Role Conflict
    David B. Booth

1983 Higher Education Research Reports

1. The Path to Excellence: Quality Assurance in Higher Education
   Laurence R. Marcus, Anita O. Leone, and Edward D. Goldberg

2. Faculty Recruitment, Retention, and Fair Employment: Obligations and Opportunities
   John S. Waggaman
3. Meeting the Challenges: Developing Faculty Careers
   Michael C. T. Brookes and Katherine L. German
4. Raising Academic Standards: A Guide to Learning Improvement
   Ruth Talbott Keimig
5. Serving Learners at a Distance: A Guide to Program Practices
   Charles E. Feasley
6. Competence, Admissions, and Articulation: Returning to the Basics in Higher Education
   Jean L. Preer
7. Public Service in Higher Education: Practices and Priorities
   Patricia H. Crosson
8. Academic Employment and Retrenchment: Judicial Review and Administrative Action
   Robert M. Hendrickson and Barbara A. Lee
9. Burnout: The New Academic Disease
   Winifred Albizu Meléndez and Rafael M. de Guzmán
10. Academic Workplace: New Demands, Heightened Tensions
    Ann E. Austin and Zelda F. Gamson

1984 Higher Education Research Reports

1. Adult Learning: State Policies and Institutional Practices
   K. Patricia Cross and Anne-Marie McCartan
2. Student Stress: Effects and Solutions
   Neal A. Whitman, David C. Spendlove, and Claire H. Clark
3. Part-time Faculty: Higher Education at a Crossroads
   Judith M. Gappa
4. Sex Discrimination Law in Higher Education: The Lessons of the Past Decade
   J. Ralph Lindgren, Patti T. Ota, Perry A. Zirkel, and Nan Van Gieson
5. Faculty Freedoms and Institutional Accountability: Interactions and Conflicts
   Steven G. Olswang and Barbara A. Lee
6. The High-Technology Connection: Academic/Industrial Cooperation for Economic Growth
   Lynn G. Johnson